Product Lifecycle Management

Edited by Fausto Pedro García Márquez

Published in London, United Kingdom

Product Lifecycle Management
http://dx.doi.org/10.5772/intechopen.1003299
Edited by Fausto Pedro García Márquez

Contributors
Alysson Helton Santos Bueno, Fausto Pedro García Márquez, Jose Antônio da Cunha Ponciano Gomes, Kourosh Parsa, Marco Antônio Sabará, Matthias Kreimeyer, Merlin Gerald Stölzle, Mohsen Naderpour, Renata Androšević, Valesko Dausch

First published in London, United Kingdom, 2025 by IntechOpen
IntechOpen is the global imprint of INTECHOPEN LIMITED, registered in England and Wales, registration number: 11086078, 167-169 Great Portland Street, London, W1W 5PF, United Kingdom

For EU product safety concerns: IN TECH d.o.o., Prolaz Marije Krucifikse Kozulić 3, 51000 Rijeka, Croatia, info@intechopen.com or visit our website at intechopen.com.

British Library Cataloguing-in-Publication Data
A catalogue record for this book is available from the British Library

Product Lifecycle Management
Edited by Fausto Pedro García Márquez
p. cm.

This title is part of the Industrial Engineering and Management Book Series, Volume 10
Topic: Production Engineering
Series Editor: Fausto Pedro Garcia Marquez
Topic Editor: Orhan Korhan

Print ISBN 978-0-85014-742-1
Online ISBN 978-0-85014-741-4
eBook (PDF) ISBN 978-0-85014-743-8
ISSN 3029-0511

If disposing of this product, please recycle the paper responsibly.

IntechOpen Book Series

Industrial Engineering and Management

Volume 10

Aims and Scope of the Series

Industrial Engineering and Management (IEM) is a discipline that focuses on optimizing complex processes and systems within various industries. It involves the integration of engineering, business, economics, mathematics, and behavioral sciences to improve efficiency, productivity, quality, and overall performance in organizations. Key aspects of Industrial Engineering and Management include: Process Optimization; System Analysis and Design; Quality Control and Management; Supply Chain Management; Operations Management; Human Factors and Ergonomics; Project Management; Cost Analysis and Financial Management; Decision Analysis.

Overall, Industrial Engineering and Management aims to optimize resources, improve processes, enhance productivity, and ensure the effective and efficient utilization of all elements involved in the production or delivery of goods and services. It is crucial in today's competitive business environment for organizations to stay efficient and competitive.

Production Engineering and Operational Excellence are fields of study and practices that focus on optimizing and improving the manufacturing and production processes within an organization. It combines principles from engineering, management, and operational strategies to enhance productivity, efficiency, quality, safety, and sustainability in the production of goods and services.

Here are the key components of Production Engineering and Operational Excellence: Process Optimization; Operational Excellence; Manufacturing Systems Design; Quality Management; Supply Chain Optimization; Production Planning and Scheduling; Automation and Technology Integration; Health, Safety, and Environmental Management; Cost Management; Performance Measurement and Key Performance Indicators (KPIs); Continuous Improvement and Innovation.

Production Engineering and Operational Excellence are crucial for organizations aiming to stay competitive in the global market by achieving high levels of efficiency, quality, and customer satisfaction while optimizing resources and minimizing waste. It is a multidisciplinary approach that encompasses engineering principles, management strategies, and the effective use of technology to drive operational success.

Meet the Series and Volume Editor

Fausto Pedro García Márquez is a Full Professor at UCLM, Spain, with accreditation since 2013. He also holds the position of Honorary Senior Research Fellow at Birmingham University, UK, and serves as a Lecturer at the Postgraduate European Institute. In addition to these roles, Fausto has experience as a Senior Manager at Accenture from 2013 to 2014. He earned his European Ph.D. with the highest distinction. Throughout his career, Fausto has received numerous awards and honors. These include the Nominate Prize (2022), Gran Maestre (2022), Grand Prize (2021), Runner Prize (2020), and Advancement Prize (2018), as well as Runner (2015), Advancement (2013), and Silver (2012) by the International Society of Management Science and Engineering Management (ICMSEM). He was also the recipient of the First International Business Ideas Competition 2017 Award. Fausto's contributions extend to academic publishing, with over 242 papers to his name. Notably, his work has been recognized in journals like "Applied Energy" (Q1, IF 9.746, Best Paper 2020) and "Renewable Energy" (Q1, IF 8.001, Best Paper 2014). His affiliations include the editorial and authorship roles in more than 50 books, with publications through respected publishers such as Elsevier, Springer, Pearson, Mc-GrawHill, IntechOpen, IGI, Marcombo, and AlfaOmega. He has authored over 100 international chapters and holds 6 patents. Fausto serves as the Editor of 5 International Journals and is a Committee Member for more than 70 International Conferences. His research portfolio encompasses being the Principal Investigator in 4 European Projects, 8 National Projects, and participating in over 150 projects involving universities and companies. His areas of expertise and research interests span Artificial Intelligence, Maintenance, Management, Renewable Energy, Transport, Advanced Analytics, and Data Science. Fausto is a recognized Expert in the European Union in AI4People (EISMD) and ESF. He also serves as the Director of www.ingeniumgroup.eu, holds the status of Senior Member at IEEE since 2021, and has been honored as an Honorary Member of the Research Council of the Indian Institute of Finance since 2021. Fausto is also the Committee Chair of The International Society for Management Science. and Engineering Management (ISMSEM) since 2020.

Contents

Preface **XI**

Chapter 1 **1**
Introductory Chapter: Product Lifecycle Management
by Fausto Pedro García Márquez

Chapter 2 **7**
Product Life Cycle
by Matthias Kreimeyer, Merlin Gerald Stölzle and Valesko Dausch

Chapter 3 **21**
Product Lifecycle and RAMS Requirement at Industry 4.0
by Kourosh Parsa and Mohsen Naderpour

Chapter 4 **41**
Enhancing Pipeline Reliability: A Structural Integrity Management Approach Using Minimal Cut Set Method and Importance Measures
by Marco Antônio Sabará, Jose Antônio da Cunha Ponciano Gomes and Alysson Helton Santos Bueno

Chapter 5 **81**
Environmental Impact of Built Structures during Their Lifecycles in Terms of Construction Waste
by Renata Androšević

Preface

Product Lifecycle Management (PLM) is a strategic approach used by businesses to oversee the entire lifespan of a product, from its conception to its disposal. It involves the management of data, processes, and collaborations across departments and stakeholders involved in product development and production. PLM aims to optimize processes, reduce time-to-market, improve product quality, and maximize profitability. Here are key aspects:

1. Centralized Data Management: PLM systems facilitate the storage, organization, and access to product-related data (designs, specifications, documents, etc.) in a centralized location. This ensures that all stakeholders work with accurate and up-to-date information.

2. Collaboration and Communication: PLM enables effective collaboration among cross-functional teams, allowing for concurrent engineering, real-time communication, and seamless sharing of information across departments and even global locations.

3. Workflow and Process Management: It involves defining and streamlining workflows, automating processes, and establishing checkpoints to ensure that each stage of the product lifecycle is efficiently managed.

4. Version Control and Traceability: PLM systems maintain version control to track changes made throughout the product development process, ensuring traceability and accountability.

5. Quality Management: Monitoring and maintaining product quality standards through the entire lifecycle, from design to manufacturing and post-sales support, ensuring adherence to regulations and customer expectations.

6. Cost and Resource Optimization: PLM helps manage costs by identifying opportunities for efficiency improvements, waste reduction, and optimal resource utilization during the product lifecycle.

7. Risk Management and Compliance: It involves assessing and mitigating risks associated with product development, production, and industry regulations and standards compliance.

8. Service and Maintenance: PLM extends beyond product development and production, covering aspects like service, maintenance, upgrades, and end-of-life strategies.

Implementing a PLM strategy requires the integration of technology, processes, and people across an organization. Companies often invest in PLM software solutions tailored to their specific industry and needs. The ultimate goal of PLM is to drive innovation, enhance competitiveness, and deliver products that meet or exceed customer expectations throughout their lifecycle.

This book is primarily designed for professionals working in diverse sectors such as production, energy, engineering, information technology, mathematics, economics, and academia. It serves as an essential resource for those seeking to deepen their knowledge of outsourcing, whether they are new to the concept or already incorporating outsourcing strategies in their daily operations. It also provides valuable insights for individuals aiming to enhance their skills in managing outsourcing processes, as well as those involved in decision-making or policy formation in this area.

The authors of this volume contribute a range of original research and practical insights, exploring both the theoretical foundations and real-world applications of outsourcing. They offer in-depth case studies that illustrate how outsourcing has been successfully implemented in various industries, showcasing its potential to drive efficiency, innovation, and cost savings. Additionally, the theoretical discussions present key concepts and frameworks that can help professionals better understand the complexities and challenges associated with outsourcing in different contexts.

Furthermore, the book delves into emerging trends and best practices in outsourcing, providing readers with the tools to navigate the evolving landscape of global business. It explores the various outsourcing models, from traditional approaches to newer, more flexible strategies, and emphasizes the importance of strategic planning and risk management in outsourcing decisions. This volume aims to equip its audience with a broad and nuanced understanding of outsourcing through a comprehensive analysis, empowering them to make informed choices and effectively implement outsourcing solutions in their organizations.

Fausto Pedro García Márquez
Ingenium Research Group,
University of Castilla-La Mancha,
Ciudad Real, Spain

Chapter 1

Introductory Chapter: Product Lifecycle Management

Fausto Pedro García Márquez

1. Understanding Product Lifecycle Management

Product Lifecycle Management (PLM) refers to the strategic process of managing the entire lifecycle of a product, from its initial conception through design, manufacturing, service, and disposal [1]. It is a systematic approach that integrates people, data, processes, and business systems to improve efficiency, innovation, and profitability across the product lifecycle. In today's competitive global markets, PLM is essential for organizations to maintain their edge, reduce time-to-market, and meet dynamic customer demands.

2. The evolution of PLM

Historically, the concept of lifecycle management emerged to address the growing complexity of product development. Before the advent of modern PLM systems, product data and processes were often fragmented across organizational silos. This led to inefficiencies, errors, and delays. Over time, advancements in information technology and the increasing need for cross-functional collaboration paved the way for comprehensive PLM solutions [2].

Modern PLM systems have evolved into highly sophisticated platforms that leverage cloud computing, artificial intelligence, and machine learning. These systems enable real-time collaboration, data analytics, and seamless integration with other enterprise systems such as Enterprise Resource Planning (ERP) and Customer Relationship Management (CRM). Additionally, the shift toward digital transformation has accelerated the adoption of PLM as businesses seek to stay competitive in a rapidly changing market.

3. Core components of PLM

The PLM framework encompasses several critical components that ensure seamless management of a product's lifecycle:

1. *Conceptualization*: The initial stage involves ideation, feasibility studies, and market analysis. Tools such as brainstorming sessions and SWOT analysis are common.

2. *Design and development*: This phase includes detailed design, prototyping, and validation. Computer-aided design (CAD) systems and digital twins play a significant role here.

3. *Production*: The manufacturing stage involves scaling designs for mass production, optimizing supply chains, and ensuring quality standards. Advanced manufacturing technologies such as 3D printing and robotics are increasingly being integrated into this phase.

4. *Operation and maintenance*: During the product's use phase, feedback from customers and performance data are collected to guide improvements. Predictive maintenance powered by IoT sensors ensures reduced downtime and extended product lifespan.

5. *End-of-life management*: Products are decommissioned, recycled, or disposed of responsibly, emphasizing sustainability. Circular economy principles are often applied to maximize resource efficiency.

4. Benefits of implementing PLM

Adopting PLM offers organizations several advantages (**Table 1**).

5. Economic analysis in Product Lifecycle Management (PLM)

Economic analysis in Product Lifecycle Management (PLM) involves evaluating the financial implications of decisions made throughout a product's lifecycle, from conception and design to disposal or recycling. This analysis is a critical component in ensuring that all stages of the product lifecycle contribute to overall profitability while maintaining efficiency and sustainability. It aims to balance costs, revenues, and risks by considering direct expenses, such as material and manufacturing costs, as well as indirect factors, including opportunity costs, market dynamics, and environmental impact [3]. Techniques such as cost-benefit analysis, net present value (NPV), return on investment

Benefit	Description
Enhanced collaboration	Facilitates cross-departmental communication and data sharing.
Reduced time-to-market	Streamlines processes, enabling faster product launches.
Cost savings	Identifies inefficiencies and reduces waste.
Improved product quality	Ensures adherence to quality standards throughout the lifecycle.
Sustainability	Promotes eco-friendly practices and compliance with the environmental regulations.
Data-driven insights	Utilizes analytics to improve decision-making and future product designs.

Table 1.
Overview of the main benefits of the economic analysis in product lifecycle management.

(ROI), and lifecycle cost analysis (LCCA) are employed to assess the economic feasibility of design choices, supply chain strategies, and market positioning [4].

By integrating financial insights into the PLM framework, companies can optimize product portfolios, minimize waste, and ensure resource allocation aligns with both short-term profitability and long-term sustainability goals [5]. This integration enables the identification of cost-saving opportunities during the early design stages, where the majority of lifecycle costs are determined, and informs pricing strategies that maximize market competitiveness. Additionally, economic analysis supports decision-making for product updates, upgrades, and discontinuation, ensuring investments in innovation and product development yield measurable returns. Ultimately, embedding robust economic analysis into PLM empowers organizations to achieve a balance between financial performance, operational efficiency, and customer satisfaction while maintaining compliance with industry regulations and sustainability standards.

6. Challenges in PLM implementation

Despite its benefits, implementing PLM is not without challenges. Common obstacles include high initial costs, resistance to change within organizations, and the need for continuous updates to technology and processes. Overcoming these challenges requires strong leadership, employee training, and a clear strategic vision [6].

A significant challenge lies in data integration, as organizations often have legacy systems that may not seamlessly integrate with modern PLM solutions. Additionally, the complexity of global supply chains and regulatory compliance adds further layers of difficulty. Organizations must also address the cultural shift required to adopt a PLM mindset, fostering collaboration across traditionally siloed departments.

7. PLM in the digital age

The integration of emerging technologies such as artificial intelligence (AI), the Internet of Things (IoT), and big data analytics has transformed PLM. Smart, connected products generate vast amounts of data that can be analyzed to derive actionable insights. This allows for predictive maintenance, real-time decision-making, and enhanced customer experiences.

Furthermore, cloud-based PLM solutions have enabled remote collaboration, making it easier for global teams to work together on complex projects. Digital thread and digital twin technologies provide a unified view of product data, ensuring consistency and accuracy across the lifecycle [7].

The advent of Industry 4.0 has further underscored the importance of PLM. As manufacturing becomes increasingly automated and data-driven, PLM serves as the backbone for managing the interplay between physical and digital systems. It also supports sustainability efforts by enabling companies to design for recyclability, reduce waste, and track environmental impact throughout the lifecycle.

8. Future trends in PLM

As technology continues to evolve, several trends are shaping the future of PLM [8]:

1. *Integration with AI and machine learning*: These technologies enhance decision-making processes, automating repetitive tasks, and providing predictive insights.

2. *Sustainability-focused PLM*: Environmental concerns are driving the adoption of tools that enable lifecycle assessments and sustainable design practices.

3. *Augmented reality (AR) and virtual reality (VR)*: These technologies are revolutionizing product design and training, providing immersive experiences for testing and visualization.

4. *Blockchain for supply chain transparency*: Blockchain technology is being explored to enhance traceability and trust within complex supply chains.

5. *Greater customization*: Advanced PLM systems are enabling mass customization by managing configurations and variations efficiently.

Author details

Fausto Pedro García Márquez
Ingenium Research Group, Universidad de Castilla-La Mancha, Spain

*Address all correspondence to: faustopedro.garcia@uclm.es

References

[1] Stark J. Product lifecycle management. In: Product Lifecycle Management: 21st Century Paradigm for Product Realisation. Germany: Springer; 2011. pp. 1-16

[2] Cholewa M. Product Lifecycle Management. Poland: Printpap Łódź; 2011

[3] Acaroğlu H, Márquez FPG. High voltage direct current systems through submarine cables for offshore wind farms: A life-cycle cost analysis with voltage source converters for bulk power transmission. Energy. 2022;**249**:123713

[4] Acaroğlu H, Marquez FPG. A life-cycle cost analysis of high voltage direct current utilization for solar energy systems: The case study in Turkey. Journal of Cleaner Production. 2022;**360**:132128

[5] Sánchez PJB, Asensio MT, Papaelias M, Márquez FPG. Life cycle assessment in autonomous marine vehicles. In: Proceedings of the Fifteenth International Conference on Management Science and Engineering Management. Vol. 2(15). China: Springer; 2021. pp. 222-233

[6] Hewett A. Product lifecycle management (PLM): Critical issues and challenges in implementation. In: Information Technology and Product Development. 2009. pp. 81-105

[7] Holligan C, Hargaden V, Papakostas N. Product lifecycle management and digital manufacturing technologies in the era of cloud computing. In: 2017 International Conference on Engineering, Technology and Innovation (ICE/ITMC). Portugal: IEEE Xplore Digital Library; 2017. pp. 909-918

[8] Ibeh CV, Awonuga KF, Okoli UI, Ike CU, Ndubuisi NL, Obaigbena A. A review of agile methodologies in product lifecycle management: Bridging theory and practice for enhanced digital technology integration. Engineering Science & Technology Journal. 2024;**5**:448-459

Chapter 2

Product Life Cycle

Matthias Kreimeyer, Merlin Gerald Stölzle and Valesko Dausch

Abstract

The product life cycle as a descriptive model of the product determines—among others—the structure and process organization within a company and has therefore a major influence on product development and its data management across a product's lifespan. In this chapter, the authors review various existing product life cycle models and recount the implications for product development. Traditional models from product development as well as business administration, technology, and circular economy are in focus. The models are contrasted in terms of their relevance and applicability in development processes. For this, aspects such as applicability to different product types or boundary conditions are examined. An outlook on IT implementation in enterprises connects this to practical applications.

Keywords: life cycle management, product life cycle, circular life cycle, life cycle models, life cycle evaluation

1. Introduction

To effectively manage complex product portfolios within a company, it is essential to have a clear understanding of the stages and states a product passes through, as well as the activities assigned to each stage. From the conceptualization of a product following the subsequent generation of a theoretical form to a physical, handleable shape created during production, different actions and associated pieces of information are required. The acquisition and description of each of these stages and states are necessary to facilitate the transition from one phase to the next. The description in models and the classification in a chronological sequence are generally referred to as the "product life cycle." The product life cycle thus provides a comprehensive description of the various forms and perspectives, which a product passes through. Depending on the specific aspects of the product life cycle, there are different life cycle models considering and covering the relevant aspects. As such, life cycle models are essential for the effective management of product-related data.

As a temporal representation of the product, the product life cycle directly impacts all parts of an organization, notably the following shown in **Figure 1**.

Following the above overview of the general concept for product life cycles, an examination of various models that have been developed for the specific description of the product life cycle is presented in the next section.

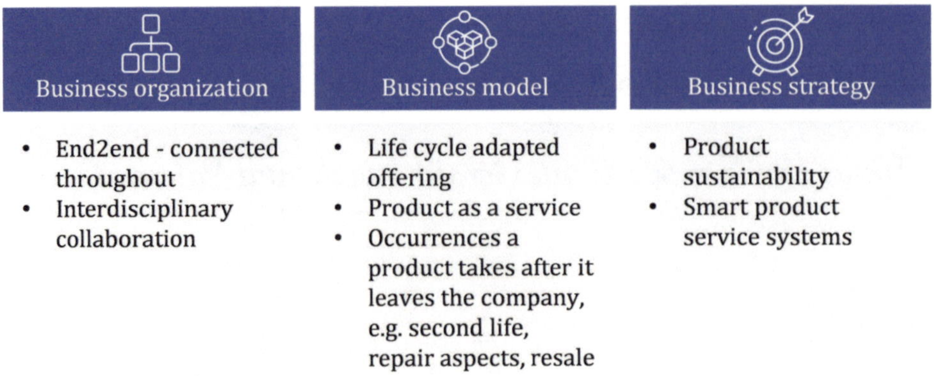

Figure 1.
Impact of the product life cycle on an organization.

2. Life cycle models

2.1 The standard life cycle model of product development

DIN ISO 15226 describes the basic structure of a life cycle model as follows: *Phases* are defined on a product-specific basis and separated from each other by *conditions*. This implies that the completion of a phase is bound to the fulfillment of a certain condition. Conversely, a phase can only be started after a certain condition has been achieved. The individual phases are delineated by *activities* that describe how the end product state can be achieved from the initial product state within a phase. These activities are assigned to *organizational units* that are responsible for their execution. This concept is illustrated in **Figure 2**.

In technical product development, the seven-phase life cycle is particularly well established and used (see **Figure 3**) [1]. The seven phases are as follows:

Requirements: In the initial phase of requirements definition, the product is described in its most abstract form. In this stage, solely the customer's demand, which the future product is designed to address, is in focus. A possible solution is developed in the form of product requirements ("What product is developed?"). These requirements consider corporate strategy, portfolio structure and the existence of similar products among competitors.

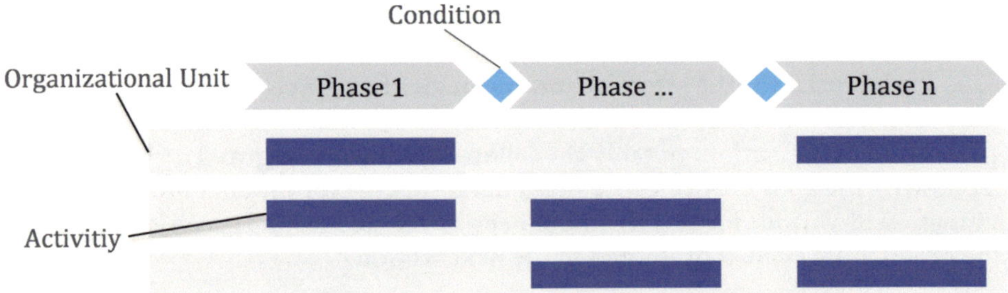

Figure 2.
General product life cycle.

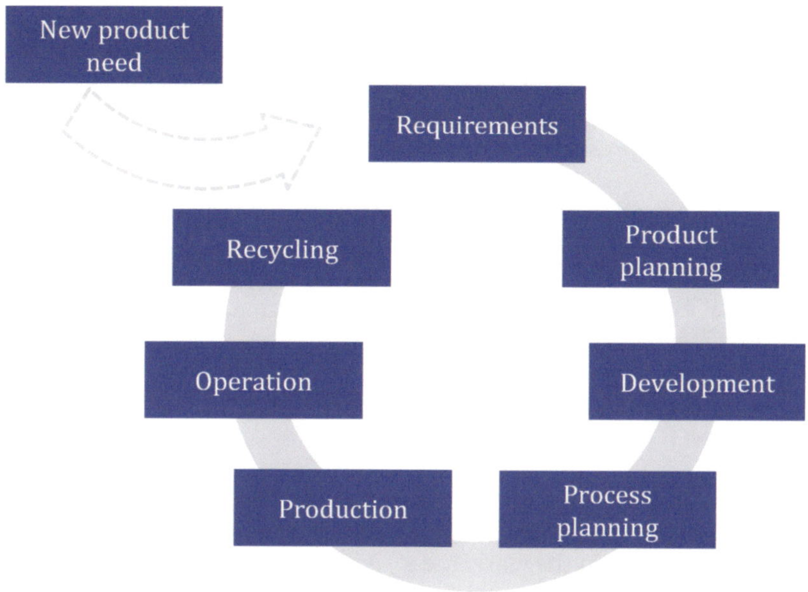

Figure 3.
Seven-phase life cycle.

Product planning: Based on the requirements, the phase of product planning begins. The primary concern is the definition of processes for the actual development of the product ("How will the product be developed?") as well as detailing the feasibility. This includes not only the determination of a project plan, the development methodology, but also financial aspects such as budgeting as well as a business case and setting up development teams. Once these aspects have been determined, the design of the product can commence.

Development: The shape and technological set-up of the product are determined in this phase. The product is designed on the basis of the information gathered from the previous phases, often involving iterations back to clarify inputs. This involves combining mechanical, electronic and IT features. The validation of the digital mock-up through simulation or testing of physical mock-ups marks the conclusion of the development phase, along with documentation (e.g., [2–4]).

Process planning: This phase determines how the product is manufactured and assembled. This includes not only the development of tools, the determination of manufacturing resources and the purchase of raw material and parts but also the final virtual and physical validation of the resulting production process.

Production: The actual manufacturing and assembly of the product occurs during the production phase, based on the set-up done before. This also includes quality assurance to persuade conformity of the product with its specification.

Operation: The transfer of ownership of the product from a company to the customer occurs during operational phase. However, the operation phase also includes services relating to the product or the maintenance and revision of the product itself during use.

Recycling: Following utilization by the customer, the product is disposed and undergoes a recycling process. Recycling can involve returning the product to production processes (refurbishing, remanufacturing), returning it as a raw material ("recycling") or thermal recovery and the resulting energy recovery.

Other life cycle models of product development typically diverge from the above approach in terms of nomenclature, the structural fineness of the phases and the level of detail in the description of the respective activities depending on the focus of the model. For purposes of completeness, the following somewhat similar models are to be mentioned, but not presented in detail [5–9].

2.2 Other life cycle models beyond standard product development

In addition to the standard product development life cycle, there are other life cycle models that are relevant, when considering different viewpoints within a company or society. The most common models are structured in accordance to business administration, technology and ecology.

2.2.1 Life cycle from a business administration perspective

The business administration product life cycle (see **Figure 4**) analyses the cash flow related to a product over the course of its development and marketing as well as the acceptance of the product in the market, e.g., with regard to volume or market share. The profit, turnover and loss of a product can be drawn from this. Many companies take such a model as a basis to calculate a business case for their products during the planning phase. Irrespective of the product, in the early stages of product development (product planning, development and preparation for market launch), companies usually incur costs that are higher than the potential revenues. Once a product is launched, sales increase and income is generated, ideally resulting in a profit. These increase during the growth and maturity phase until a saturation point is reached, leading to declining sales and therefore declining profits, which may be so low in the decline phase that a product can no longer be manufactured profitably. The length of the phases, the rise and fall of profits and sales, and the level of costs are highly variable and depend substantially on the product, the technologies used and other market conditions. The business administration product life cycle helps companies to analyze and control products on a cost basis in order to make strategic decisions. It is, in fact, essential, therefore, to understand the life cycle phases to plan for a successful product across its life [10].

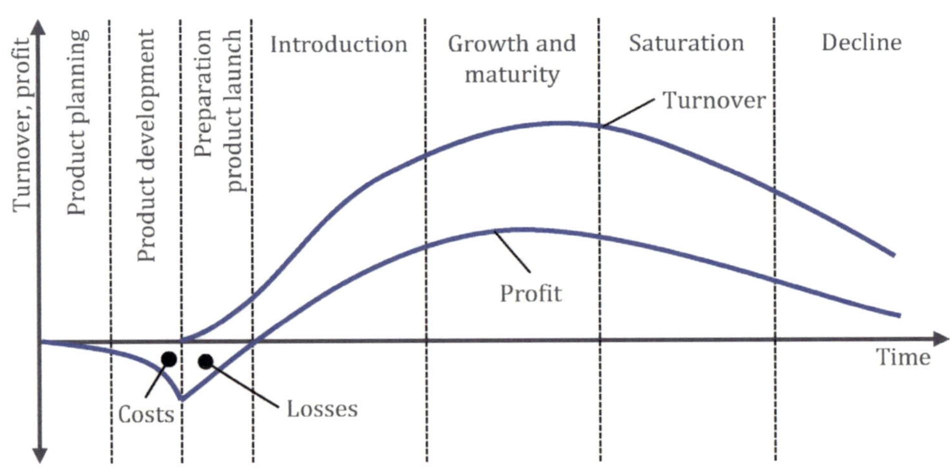

Figure 4.
Life cycle from a business administration perspective based on Ref. [10].

2.2.2 Technology life cycle

The technology life cycle describes not just a single product, but fundamental technologies that can be used in products either alone or in combination. It, therefore, is somewhat similar to the business life cycle: After its launch, a technology only starts slowly; early adopters use it in isolated cases. As the technology moves into the growth phase, it increasingly becomes more accepted or mature, before gradually moving to saturation in a maturity phase. At some point then, the degeneration phase begins, as new technologies emerge and take over, and existing ones become obsolete. For companies, the time of adaptation and evolution of a technology in products is crucial for the competitiveness and economic position of the company. The analysis of technology life cycles therefore directly contributes to innovation management in companies. The technology life cycle is more of a descriptive approach to technology, for example as a function of dissemination (**Figure 5** left) [11] or performance (**Figure 5** right) of a technology [12]. However, the technology life cycle is only indirectly related to the product through the technology used in the product.

It should be noted that technology life cycles are generally longer than the corresponding product life cycles from a business administration perspective (see **Figure 6**). The technology matures over several successive product generations until it becomes obsolete and is replaced by a new technology. The evolution of internal combustion engines (ICEs) in cars provides an illustrative example of technological advancement. Over the past decades, ICEs have undergone significant improvements in various aspects, such as emissions and performance. These generations of ICEs have spanned multiple product life cycles. Currently, the overall technology of ICEs is being superseded by battery electric vehicles. This technology is also being enhanced with each new product generation, as evidenced by the increased range due to advancements in battery technology.

2.2.3 Ecology perspective on the product life cycle

The ecology product life cycle (see **Figure 7**) is also known as the material life cycle. This approach places the material used at the center of attention. The focus lies on the environmental impact at each stage of the life cycle. The material includes not only the actual product material but also the resources and transformation processes and emissions. It considers all phases of the life cycle, from extraction of

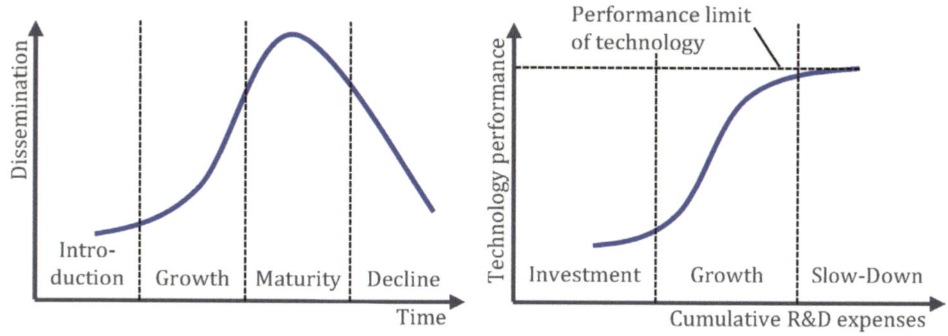

Figure 5.
Technology life cycle.

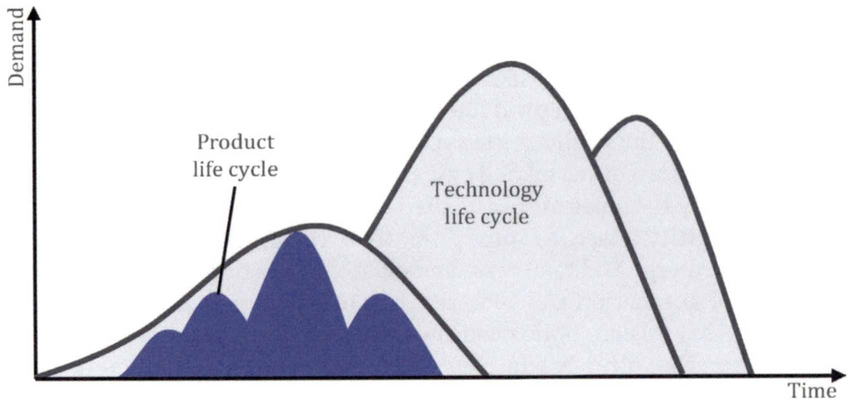

Figure 6.
Comparison of technology and product life cycles.

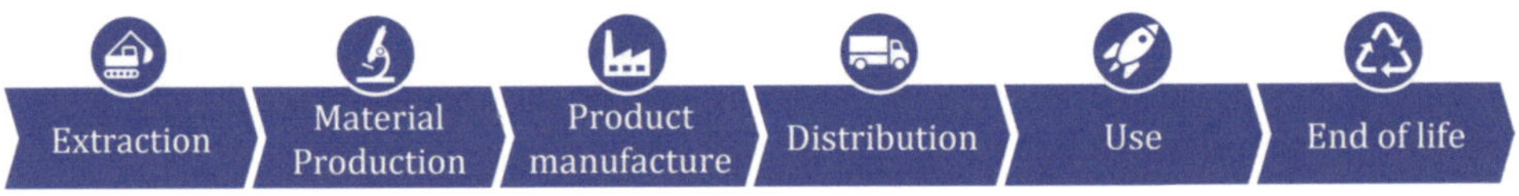

Figure 7.
Ecology perspective on the product life cycle.

raw materials, through processing, to reuse or disposal at the end of life. The ecological approach to product life forms the basis for life cycle costing (LCC) methods, life cycle assessments and social aspects [13].

The growing need for sustainability in the fourth industrial revolution implies that the ecological consideration and assessment of products are becoming increasingly important. Companies can actively promote the sustainable design and development of products to meet important customer demands.

2.3 Implementation of life cycles in companies: Adapting the model

The previous chapter illustrates that there are various viewpoints on a (product) life cycle for different purposes within companies. Due to increasing complexity in product development as well as changing requirements from customers and society alongside technological progress, it becomes more and more important for companies to enable a comprehensive view on the manifold perspectives of the (product) life cycle. Hence, the implementation of life cycle models depends on both internal and external boundary conditions, such as the product itself, the company strategy or demands from the market, especially its business model. In other words, life cycle models require adaption to the specific setting they are used in. Consequently, a generic "one-size-fits-all" life cycle model is not practically feasible. Especially for complex or new products, it therefore makes sense to have roles in place within a company entrusted with shaping and managing life cycles. Often, this is done in the product management department.

Again, the transformation from ICE to battery electric vehicles servers as an illustrative example: while in the past, the vehicle including the combustion engine was well understood and underwent a very classic process of spare parts when needed and usually a first life and then a second life as a used car, this life cycle is much less clear for an electric vehicle, with the battery being the most expensive single component and the uncertainty of how long it will live; in turn, today, we see a lot of uncertainty around used electric vehicles, as the state of the battery often is unclear. With this in mind, it is quite obvious that in the long run, electric vehicles will likely need a different life cycle model.

2.4 Implementation of life cycles in companies: IT tools to support the model

Product life cycle management (PLM) systems have been developed since the 1990s to support product documentation. PLM is, in this context, defined as a comprehensive and integrated management and organizational approach that unifies and oversees all information pertaining to a product's life cycle. The evolution of PLM systems has been driven by the need to effectively manage the growing complexity and diversity of variant-rich product portfolios. In conjunction with the advent of virtual development and technological advancement, it became imperative to provide support for this through the utilization of information technology. Even though the concept of PLM clearly indicates that all information pertaining to a product throughout its life cycle must be managed, in many instances, PLM systems are employed solely for the management of computer-aided design (CAD) data. More recently, PLM systems have evolved to encompass a broader scope, including mechatronics in products and the concomitant emergence of cyber-physical systems (CPS), along with the mounting significance of sustainability, which are collectively transforming the very nature of the product life cycle. As a consequence, products are nowadays more capable of undergoing more than one life cycle. This is also the case with regard to the software incorporated into products. Smartphones serve as a representative case in point. Hardware is replaced on an annual basis, whereas software is updated on a much more frequent basis [1, 14, 15].

In order to accommodate such changes, networked and graph-based approaches are now employed to facilitate the most comprehensive view possible of all domains and disciplines involved in a life cycle. This signifies that all potential data is accessible in a centralized repository at all times. Subsequently, the requisite views of the model are derived for the application [16].

There is extensive literature on this field that cannot be detailed here. For further details, see for example [7, 16, 17].

3. Decision support for companies

A set of key performance indicators (KPIs) is usually a core instrument to assure a company's operational and strategic success. KPIs facilitate the assessment of diverse perspectives on the life cycle. The following section outlines methodologies for determining various KPIs with regard to the product's life cycle.

3.1 The economic assessment

LCC is a comprehensive methodology that evaluates the success of a product or service over its entire life cycle, considering both technical and organizational factors. The objective of LCC is to maximize the overall benefit derived from a product over its entire life cycle. It also compares the expenses and income in all phases of the product life cycle. This approach of a permanent life cycle-spanning profit and loss account is intended to enable predictions of the economic effects of technical and organizational measures within the framework of successful technology management.

A review of product life cycles reveals that there are frequently temporal discrepancies between the initial determination of costs and revenues and their subsequent development over the product's lifecycle. This is due to the trade-offs involved in the decision-making process. LCC can be employed to identify these interactions and make them available for future-oriented and long-term cost and revenue management. By connecting technical and operational data from previous life cycles with prospective sales and technology forecasts, new possibilities can be identified, enabling the efficient operation of the plant at its limits of performance and precision. The possibility of "system upgrading" represents a particularly crucial function. An LCC analysis can be employed to ascertain the optimal time for refurbishment from a business perspective and to forecast the company's profitability. In this manner, the refurbished plant enters a second life cycle, thereby exerting a significant positive influence on the overall economic success.

Another method to assess the product with its economic performance in mind is the total cost of ownership (TCO), which is commonly used. It follows the products' life cycle and collects costs that occur "due to the use of a product," such as the need for electricity for a plant or the salary of a driver for a truck. Hence, it provides a very comprehensive view on the product life cycle, particularly from the customer's or user's point of view.

3.2 Environmental assessment

In recent years, the environmental footprint of products has become increasingly important for product development [18, 19]. In addition to purely economic aspects, potential environmental impacts affect the company's sustainable business operations in the form of regulations.

Life cycle assessment (LCA) is a methodology for evaluating the environmental impact of a product, process or service throughout its entire life cycle, from raw material extraction to disposal. LCA records and evaluates the environmental impact of products, processes and services over the entire life cycle, from the initial extraction of raw materials to the eventual disposal of the product. At its heart, it models the actual life cycle and then attributes values to it. With this, an LCA allows for the identification of weaknesses in an ecological and economic context. To conduct such an assessment, it is essential to include comparative products in the assessment. This comparison is typically conducted following the international standardization status for ecological balancing (e.g., DIN EN ISO 14040), thereby establishing the basis for communicable and widely accepted statements regarding the ecological improvements achieved through the new design solution.

3.3 Integration of economic and ecological evaluation in the product development process

By carrying out LCC, TCO and LCA at an early stage, economic and ecological requirements can be specifically addressed during product development. In this way, not only can the actual values of complete products be determined, but also concept decisions can be made during product development. It should be noted here that the validity of the methods decreases due to the abundance of data available as a basis for calculation. The earlier these methods are used in the product development process, the less accurate the results will be (due to estimates and failure to consider aspects that may be relevant later).

4. Changes and challenges in life cycle models

It can be observed across all the modes that the models do not allow for iterations within individual or between several phases in the general sequential process. The work packages are processed in a linear manner. Nevertheless, the reality of product development is driven by uncertainties, thus needing feedback loops and iterations [20].

A further point of convergence among the various product life cycle models is the clear delineation of the individual phases from an organizational viewpoint. This leads to the need of robust interface definitions within the company in order to ensure the seamless transfer of information. Often, in real life, these are less clear, and the approach a company chooses needs to be very clearly communicated internally. This is even more important as organizations become more and more agile these days, leading to frequent adjustments in the interfaces between departments.

The increasing prevalence of CPS and integrated services introduces an additional layer of complexity to the alignment of the life cycles of different domains, given the variability in their respective phases. Consider, for example, the life cycles of mechanical components that are typically considerably longer than those of software components. The alignment of these different cycles with different durations represents a significant challenge. The example of Apple's AirPods recent functional update illustrates that products may exhibit entirely distinct life cycles for their hardware and software components and their functions, as functions get updated and substantially adapted over time while hardware remains the same.

In recent years, there has been a notable increase in the relevance of sustainability as a purchasing consideration for customers. As a result of the implementation of sustainability strategies, products are now capable of undergoing multiple life cycles. This also presents a challenge for established models and their associated methodologies, as the concept of consecutive life cycles was not previously considered. Furthermore, the field of sustainability requires the collection and analysis of data, which was previously not a concern for companies. The retrieval and management of such data therefore represents a significant challenge for companies.

5. From product life cycle to circular economy

So far, it was assumed that the product life cycle has a beginning ("cradle") and an end point ("grave"). This may have been true at the beginning of product life cycle modeling. However, these days there is a transformation from a linear to a circular economy. Products should not simply be disposed of after use but should serve as raw material for future products. "Cradle to grave" is now becoming "cradle to cradle" [21]—that is, the repeated passage through the life cycle, making it circular. Entire products, components or just the raw materials can be returned to a new life cycle. It is important to note that this circularity has an impact on the other life cycle models. The linear economy assumes an almost complete loss of value of the product during

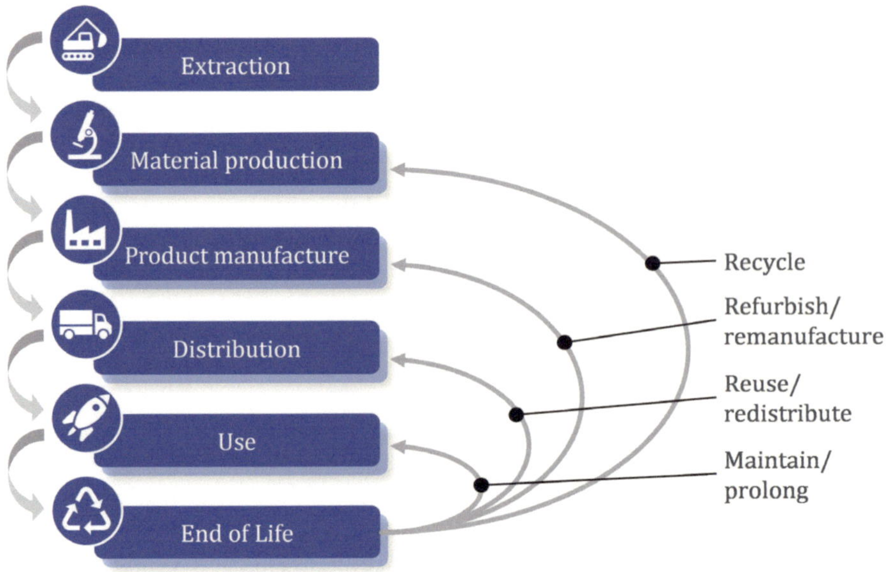

Figure 8.
R-strategies.

use [13]. This is not the case with a return into the life cycle, through one of the so-called R-strategies (see **Figure 8**). These economic aspects also need to be considered at the beginning of a product's life cycle even more so in the business model orientation and calculation. Resource consumption and emission assessments of a product in a holistic balance within a circular economy also change significantly in contrast to a linear product life cycle.

The potential for the product to be returned, either in its entirety or in part, is significantly influenced by the product design itself (e.g., the ability to dismantle the product) and therefore must be considered during the development process. To optimize the utilization of a given product, a forward-thinking approach to product life cycle is needed to incorporate the additional aspects from it. Upon the conclusion of the utilization phase, the operator of the product is confronted with two potential avenues: disposal or recycling of the product, or a so-called product upgrade. An upgrade can be defined as a transformation of the product into a novel operational state, as evidenced by the introduction of new product functions. Modifications to the product are made in the form of software or hardware changes, which serve to enhance, extend or introduce new functional features relative to the product's original state. Similarly, the product can be enhanced, expanded or repurposed for applications that are entirely novel. The selection of an upgrade effectively marks the commencement of a new phase of the product's existence. Nevertheless, the possibility of an upgrade is not always technically feasible or economically viable.

From an organizational perspective, data collection can facilitate the process of optimization. The utilization of data logging enables the generation of statistical analyses pertaining to the behavior of products, the monitoring of products or processes in an online capacity and the real-time observation of product performance. The data thus obtained is then subjected to analysis in accordance with pre-established criteria, thereby revealing potential for optimization. This approach allows for the effective monitoring of machinery, enabling not only technical optimization but also the conduct of profitability and economic efficiency analyses, as well as forward-looking planning based on the utilization of authentic machine data.

6. Conclusion

This chapter introduced the product life cycle as a descriptive model and its relevance to companies. First, the concept of a life cycle and its implications for different parts of a company were examined. Then, the standard elements of the standard life cycle according to DIN ISO 15226 and a widely used version of this standard life cycle were presented. This standard model is then contrasted with other models that focus on other perspectives of a product, the company and society, such as business administration, technology and ecology. With the continuous changes in product development, it becomes more and more relevant that a comprehensive approach is used within companies to consider all models and their perspectives in order to remain competitive and to support strategic decisions for development. It is therefore necessary to actively manage the implementation of life cycle models in an organization. This requires an appropriate organizational set-up as well as a proper IT implementation. In a nutshell, from an organizational point of view, it is necessary to have roles that are responsible for managing the life cycle of products and technologies in the company. These roles should also be actively involved in strategic and operational product decisions. For the IT implementation, it is crucial to take advantage of the latest advances in current PLM systems, which strive to go beyond the capabilities of being a CAD management tool to also support other life cycle phases, while also supporting changes in products, such as the ability to manage multiple life cycles.

To support decision-making, three key performance indicators have been introduced that focus on different aspects of the life cycle models. LCC and TCO address the growing need for a comprehensive view of products to further improve business revenue and customer value over the entire life cycle. In addition, LCA addresses the growing importance of sustainability in today's products. It enables companies to assess the environmental impact of their products.

Current challenges for life cycle models include adapting from a strictly linear model to a more iterative and agile model in terms of working methods and changing organizational structures within companies. Added to this is the creation of an increasing number of CPS with integrated services and the general trend toward a sustainable economy.

This last trend is so relevant that it is drastically changing the way products are developed, used and handled at the end of their life.

Conflict of interest

The authors declare no conflict of interest.

List of abbreviations

ICE	internal combustion engine
PLM	product life cycle management

CAD	computer-aided design
CPS	cyber-physical systems
KPI	key performance indicator
LCC	life cycle costing
TCO	total cost of ownership
LCA	life cycle assessment
IT	information technology

Author details

Matthias Kreimeyer*, Merlin Gerald Stölzle and Valesko Dausch
University of Stuttgart, Institute for Engineering Design and Industrial Design, Stuttgart, Germany

*Address all correspondence to: matthias.kreimeyer@iktd.uni-stuttgart.de

References

[1] Eigner M, Stelzer RH. Product-Lifecycle-Management: Ein Leitfaden für Product Development und Life-cycle-Management. 2nd ed. Berlin, Heidelberg: Springer; 2009

[2] Association of German Engineers. VDI 4500: Technische Dokumentation. Berlin, Germany: Beuth Verlag GmBH; 2006

[3] German Institute for Standardization. DIN 6789: Dokumentationssystematik—Verfälschungssicherheit und Qualitätskriterien für die Freigabe digitaler Produktdaten. Berlin, Germany: Beuth Verlag GmBH; 2013

[4] German Institute for Standardization. DIN EN 82045-1: Dokumentenmanagement Teil 1: Prinzipien und Methoden. Berlin, Germany: Beuth Verlag GmBH; 2001

[5] Gericke K, Bender B, Pahl G, Beitz W, Feldhusen J, Grote K-H. In: Bender B, Gericke K, editors. Der Produktentwicklungsprozess, Pahl/Beitz Konstruktionslehre. Berlin, Heidelberg: Springer Berlin Heidelberg; 2021

[6] Grieves M. Product Lifecycle Management: Driving the Next Generation of Lean Thinking. New York: Mc-Graw-Hill; 2006

[7] Stark J. Product Lifecycle Management (Volume 1): 21st Century Paradigm for Product Realisation. 4th ed. Cham: Springer International Publishing AG; 2020

[8] Association of German Engineers. VDI 2219: Informationsverarbeitung in der Produktentwicklung: Einführung und Betrieb von PDM-Systemen. Berlin: Beuth Verlag; 2016

[9] Association of German Engineers. VDI 2221: Entwicklung Technischer Produkte und Systeme Modell der Produktentwicklung. Berlin: Beuth Verlag GmBH; 2019

[10] Grote K-H, Engelmann F, Herbst S, Beitz W. Grundlagen der Produktentwicklung. In: Hennecke M, Skrotzki B, editors. Hütte. 35th ed. Berlin, Heidelberg: Springer Vieweg; 2022. pp. 575-591

[11] Gochermann J. How do technologies change over time? In: Gochermann J, editor. Technology Management. Wiesbaden: Springer Fachmedien Wiesbaden; 2022. pp. 5-7

[12] Foster RN. Innovation: Die Technologische Offensive. Wiesbaden: Gabler Verlag; 1986

[13] Liu L, Ramakrishna S, editors. An Introduction to Circular Economy. Singapore: Springer Singapore; 2021

[14] Vogel-Heuser B, Fantuzzi C, Wimmer M, Böhm M, Fay A. Herausforderungeninderinterdisziplinären Entwicklung von Cyber-Physischen Produktionssystemen. Automatisierungstechnik. 2019;**67**(6):445-454. DOI: 10.1515/auto-2018-0144

[15] Zehbold C. Product lifecycle management (PLM) im Kontext von Industrie 4.0. In: Hofmann J, Fend L, editors. Digitalisierung in Industrie-, Handels- und Dienstleistungsunternehmen. Wiesbaden: Gabler; 2018

[16] Eigner M. System Lifecycle Management: Engineering Digitalization (Engineering 4.0). Wiesbaden, Heidelberg: Springer Vieweg; 2021

[17] Watts FB. Engineering Documentation Control Handbook: Configuration Management and Product Lifecycle Management. 4th ed. Norwich, NY, London: Elsevier Health Sciences Distributor; 2012

[18] Schiessl A, Müller R, Volk R, Zimmer K, Breun P, Schultmann F. Integrating site-specific environmental impact assessment in supplier selection: Exemplary application to steel procurement. Journal of Business Economics. 2020;**90**(9):1409-1457. DOI: 10.1007/S11573-020-00967-1

[19] He B, Xiao J, Deng Z. Product design evaluation for product environmental footprint. Journal of Cleaner Production. 2018;**172**:3066-3080. DOI: 10.1016/J.JCLEPRO.2017.11.104

[20] Wallace K, Pahl G, Beitz W, Feldhusen J, Grote K-H, Blessing LTM, editors. Engineering Design: A Systematic Approach. 3rd ed. London: Springer; 2007

[21] Braungart M, McDonough W. Cradle to Cradle: Einfach Intelligent Produzieren. Munich: Piper ebooks; 2014

Chapter 3

Product Lifecycle and RAMS Requirement at Industry 4.0

Kourosh Parsa and Mohsen Naderpour

Abstract

Product Lifecycle Management and Reliability, Availability, Maintainability, and Safety are crucial in product development. This chapter explores their integration for enhanced product performance and customer satisfaction, emphasizing the critical role of reliability, availability, maintainability, and safety (RAMS) within Product Lifecycle Management (PLM). Integrating RAMS considerations into the PLM process massively enhances product value, resulting in more robust products and reduced long-term service costs. This synergy drives innovation and sustains competitive advantage in the marketplace. Real-world instances across industries demonstrate the value of combining these approaches to enhance product development and performance. This research stresses the importance of integrating PLM and RAMS to achieve a comprehensive view of product development, ensuring efficient and safe lifecycle management.

Keywords: RAMS, PLM, functional security, reliability, availability, maintenance, safety, software management, agility and sustainability

1. Introduction

Industry 4.0 represents the fourth industrial revolution, where the virtual and physical worlds merge through technologies such as Cyber-Physical Systems, the Internet of Things (IoT), and Smart Factories. PLM is a strategic method that oversees every phase of a product's lifecycle, from conception to disposal, covering design, manufacturing, maintenance, and end-of-life stages. PLM acts as a facilitator for Industry 4.0. Digital platforms like Enterprise Resource Planning (ERP) utilize IoT, simulation, and additive manufacturing.

PLM serves as a vital link between Industry 4.0 technologies and enterprise asset management. At the same time, reliability, availability, maintainability, and safety (RAMS) focuses on ensuring product reliability, minimizing downtime, and enhancing safety [1]. PLM supports RAMS by enabling design for reliability and maintainability, managing product changes and updates, ensuring traceability and compliance with safety standards, and integrating RAMS data into the product lifecycle. Therefore, PLM plays a crucial role in Industry 4.0 and RAMS, fostering innovation, efficiency, and safety throughout a product's journey.

PLM software is tailored to oversee data and processes throughout a product's entire lifecycle, from inception to launch. This software aims to standardize and automate tasks, streamline product development, and optimize resource allocation. PLM has evolved to encompass a more comprehensive concept in recent decades. Leading manufacturing organizations have expanded their systems and methodologies to enhance product lifecycle coverage and traceability. The emergence of Industry 4.0 presents challenges, such as the development of intelligent products and their integration through digital models. The future of production envisions mass customization and flexible production systems, requiring investment in software and cloud-based services to create smart products that can be customized and remotely controlled using IoT technologies. The concept of Industry 4.0 entails fully digitized production, incorporating features such as interoperability, virtualization, decentralization, real-time capability, service orientation, and modularity. In advanced smart factories, CPS processes are monitored using virtual mapping and decentralized decision-making. With IoT support, real-time communication and collaboration occur among individuals, machines, or between individuals and machines. Process engineers now use various software tools to achieve solutions quickly. However, challenges persist, especially with engineers collaborating on large projects. Industry 4.0 necessitates a modular and flexible factory structure, and its success relies on the consistent development and management of an agile methodology [2].

The key benefits of PLM include improved collaboration among cross-functional teams, seamless integration of various areas into a comprehensive product development strategy, enhanced product quality through process standardization, accelerated new product introductions, and seamless integration with computer-aided design (CAD) systems, as shown in **Figure 1**. PLM ensures efficient software development and the delivery of high-quality products by bridging the gap between innovation, development, and commercialization [3]. The goal is to elaborate a methodology that can meet Industry 4.0 requirements, accommodate standard modeling techniques, and provide a unified framework for the product lifecycle design, implementation, integration, and management.

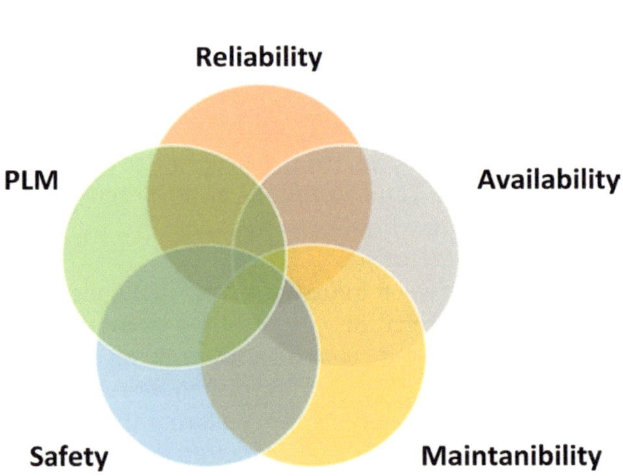

Figure 1.
PLM vs. RAMS.

2. PLM and RAMS requirements

In the realm of engineering and product development, ensuring the RAMS of intricate systems holds paramount importance. These factors directly influence overall performance, cost-effectiveness, and user satisfaction. This section aims to delve into the importance of RAMS requirements within the context of PLM:

- Reliability pertains to a system or component's capacity to operate without failure over a specified duration. Engineers evaluate reliability using metrics such as Mean Time Between Failures (MTBF) and Failure Rate (λ). Understanding the dependability of individual components and the overall system is crucial for optimizing design and mitigating risks in PLM [4, 5].

- Availability gauges a system's readiness for use as required, considering maintenance downtime, repair time, and preventive maintenance schedules. PLM tools assist in modeling availability scenarios, aiding engineers in optimizing maintenance strategies and minimizing downtime [1].

- Maintainability focuses on the ease of maintenance, repair, and replacement, considering factors such as Mean Time To Repair (MTTR), component accessibility, and documentation. PLM systems facilitate tracking maintenance data, predicting maintenance needs, and ensuring efficient serviceability [4, 5].

- Safety encompasses accident prevention, risk minimization, and compliance with regulations. PLM platforms integrate safety analysis tools, such as Failure Modes and Effects Analysis (FMEA) and Fault Tree Analysis (FTA), to identify potential failure modes, their consequences, and critical components (**Figure 2**) [6].

RAMS requirements ensure engineering systems' robustness, safety, and efficiency. Engineers can use PLM platforms and specialized tools to make informed

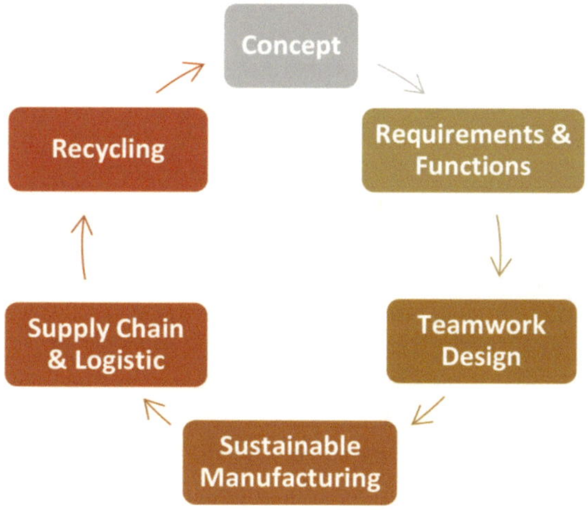

Figure 2.
Product lifecycle stages.

decisions, optimize designs, and enhance system performance. Organizations can deliver reliable and sustainable solutions by integrating RAMS considerations early in the product lifecycle [4].

2.1 The product lifecycle stages

The product lifecycle encompasses the stages a product passes through, from its inception to decline. Understanding this lifecycle is crucial for businesses to make well-informed decisions. Additionally, RAMS serve as essential metrics in engineering and product management, ensuring that a product meets operational standards throughout its lifecycle [7]. The correlation between the product lifecycle and RAMS is emphasized, emphasizing the potential for integrating RAMS principles to enhance product success at each stage (**Figure 3**).

- *Introduction stage*: Products are introduced, and awareness is developed. Sales experience gradual growth as early adopters gauge the market. RAMS implications during this stage, it is crucial to prioritize safety and maintainability. Initial designs should emphasize reliability to introduce confidence among early users. Rigorous testing protocols are necessary to identify potential failures before a wider release.

- *Growth stage*: During this phase, sales significantly increase as the product gains acceptance in the market. Competitive actions also escalate. In terms of RAMS implications, it is crucial to emphasize reliability and availability. Customers expect consistent performance, and any failure can undermine trust. It is important to prioritize investments in scalable manufacturing processes and maintain quality standards to meet the growing demand.

- *Maturity stage*: As sales growth stabilizes and the market becomes saturated, competition intensifies, resulting in price wars and the adoption of marketing strategies to set products apart. There is a shift toward prioritizing maintainability and operational performance (RAMS implication). The focus is on the belief that a well-maintained product can offer a competitive advantage. Consequently, establishing feedback systems to capture data on product performance becomes crucial, allowing for updates or modifications that enhance reliability and customer satisfaction.

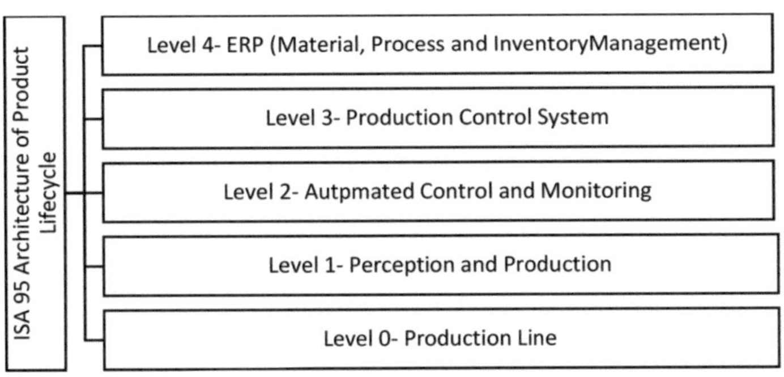

Figure 3.
Product lifecycle stages [8].

- *Decline stage*: As sales decline due to market saturation, technological advancements, or changing consumer preferences, businesses are confronted with critical decisions. They must ascertain whether to invest in rejuvenating the product or discontinuing it. Maintaining safety standards as production decreases is vital, focusing on reducing production without compromising safety or reliability.

Engineering Change Management is crucial in efficiently processing change and reducing manufacturing waste across product development, production, quality, cost, and delivery stages [9].

2.2 The significance of RAMS in PLM

Incorporating RAMS into product lifecycle management offers several key advantages:

- Enhanced customer trust: Businesses can build user trust by emphasizing reliability, availability, maintainability, and safety. A positive reputation for quality can lead to increased customer loyalty and repeat purchases.

- The implementation of RAMS can significantly lower operational and maintenance costs. A well-designed and reliable product necessitates less frequent repairs, thereby reducing downtime and ultimately resulting in cost savings [10].

- Due to stringent safety and performance standards, regulatory compliance is paramount across numerous industries, mainly transportation, healthcare, and manufacturing. Adhering to RAMS principles mitigates legal risks and bolsters product credibility [11].

- Input from RAMS analysis can catalyze product innovation, leading to enhancements in existing products or the inspiration for novel solutions. Establishing an iterative feedback loop encompassing product performance data and various lifecycle stages fosters a climate of continuous improvement [2].

2.3 Integrating RAMS in PLM

Effective collaboration among design, engineering, manufacturing, and marketing functions enhances RAMS. This collaboration ensures that all aspects of product development are aligned with RAMS goals from the outset. Implementing simulations, prototypes, and field testing can identify potential RAMS-related issues early in the product lifecycle. This iterative approach guarantees that reliability and safety are integrated into the product prior to mass production [1]. Leveraging analytics and data collection tools throughout the product lifecycle allows businesses to monitor RAMS metrics actively. This data informs decisions related to maintenance schedules, design changes, and customer support strategies. Following a product's launch, continuous monitoring for reliability and safety is essential to gather insights for potential product adjustments or future designs. This practice ensures sustained alignment with customer requirements and adherence to safety standards.

Integrating RAMS principles throughout all stages of the product lifecycle is crucial to ensuring product success and safety. This fosters a culture of quality, enhances customer trust, and creates greater long-term value for businesses. Maintaining high

RAMS standards becomes increasingly important as product complexity increases, making RAMS a vital focus for future product development strategies. Therefore, understanding and effectively managing this relationship is critical for companies striving to thrive in competitive markets.

3. PLM and functional safety

This section explores the vital link between the product lifecycle and functional safety in engineering disciplines such as automotive, aerospace, medical devices, and industrial systems. It emphasizes the importance of integrating functional safety throughout each stage, ensuring compliance with safety standards while optimizing product performance and reliability. It is imperative to meticulously articulate the RAMS requirements in line with the product characteristics. Regarding safety systems, it is essential to refer to IEC 61508 [4], IEC 61511 [5] and IEC 62061 [12] to ascertain supplementary requirements pertaining to hardware and software characteristics, particularly architectural constraints, diagnostic capabilities, and behavior in fault conditions. In addition, it is crucial to formulate an initial product description encompassing the system, subsystems, and components. Product safety requirements are about the safety performance of products, as well as all intended uses and foreseeable misuses. These requirements are primarily delineated in laws, regulations, and standards or specified by customers of specialized products [1, 11].

In today's technology-driven world, products must meet performance criteria and adhere to stringent safety standards. Therefore, it is essential to understand how to incorporate functional safety into the product lifecycle to deliver secure products to the market.

- Concept development involves identifying market needs, generating ideas, and creating initial design concepts. Conducting an early assessment of safety requirements and hazards is crucial to ensure functional safety. Initial risk assessments should be carried out using methodologies such as the Hazard and Operability Study (HAZOP) to identify potential safety issues and define safety goals for the system.

- Design and development: This stage encompasses detailed design work, prototyping, and product development. At this juncture, adhering to functional safety standards such as ISO 26262 [13] for automotive systems or IEC 61508 [4] for general safety in control systems is crucial. Design practices should encompass fault tolerance and redundancy mechanisms to ensure the system can effectively handle various failure modes.

- Testing and validation: Thorough testing, verification, and validation of designs before market introduction are imperative. Functional safety testing must ensure the system operates as intended under all operational conditions. This encompasses functional tests, fault injection tests, and safety audits. Techniques such as FMEA should be iterated to verify assumptions made in earlier phases.

- Market introduction: This phase involves launching the product and increasing production. It is important to monitor for unexpected safety issues through user feedback systems. Conducting a safety case review and ensuring that all regulatory requirements are met before the launch are crucial elements during this phase.

- Production and distribution: This encompasses mass production and distribution of the product to consumers. Quality control processes and strict adherence to safety procedures during manufacturing can help mitigate defects that could compromise functional safety. Establishing a framework for continuous reporting and safety assessment is crucial.

- Post-launch monitoring entails the critical task of tracking product performance and soliciting customer feedback in the market. It involves implementing field data analysis to detect any safety-related issues following the product launch. By establishing a robust feedback loop, manufacturers can promptly address emerging safety concerns, thereby ensuring continuous compliance with functional safety requirements.

- Discontinuation and disposal: Gradually stop using the product and properly dispose of it. It is crucial to plan for the safe disposal and management of any potential hazards associated with the product at the end of its life. This planning should encompass strategies for handling hazardous materials and ensuring that systems at the end of their lives do not pose safety risks.

3.1 PLM and software safety, ensuring reliable software products

Traditionally, products meant physical items for sale. Over decades, the term evolved to include both tangible and intangible items. Tangible products include raw materials, goods, or assets; intangible ones include systems, services, applications, virtual products, or intellectual property [14]. PLM is a comprehensive process that oversees the entire lifecycle of a product, from its conception and design through manufacturing to service and disposal. In the context of software development, PLM is crucial for ensuring that software products are not only functional but also safe and reliable. Software safety, which involves ensuring that software operates without causing harm to users, property, or the environment, is a critical aspect of this lifecycle. As software systems become more complex and integrated into critical infrastructure and everyday products, the relationship between PLM and software safety becomes increasingly important. PLM integrates people, processes, systems, and information to manage a product's entire lifecycle, including software development stages:

- During the initial phase, it is imperative to delineate the product's underlying purpose, pinpoint market needs, and ascertain technical and financial feasibility. Moreover, this stage necessitates identifying safety requirements to guide the ensuing design and development processes [15].

- Detailed design specifications are created in the Design and Development phase, and the software is developed. Safety must be considered early in the design process to prevent potential risks. Techniques such as fault tree analysis, hazard analysis, and safety-critical design practices are often utilized.

- Upon completion of software development, the product undergoes meticulous testing to verify its adherence to safety and performance standards. Testing methodologies, including static code analysis, dynamic testing, and formal verification, are employed to identify and rectify safety concerns [16].

- Following the software's deployment, a continuous monitoring and maintenance regimen is essential to guarantee sustained safety and reliability. This entails implementing routine updates, patches, and safety evaluations to address emerging vulnerabilities and evolving user requirements effectively [16].

- End-of-Life: Software products inevitably reach the end of their operational lifespan and necessitate decommissioning. Methodical management during this phase is imperative to guarantee the secure retirement of software, thus preventing the persistence of unresolved vulnerabilities or unaddressed safety concerns [17].

3.2 The importance of software safety

The Software Development Lifecycle Management (SDLC) encompasses software product planning, creation, testing, and deployment. Diverse methodologies are employed, such as the spiral model, the waterfall model, and the agile methodology. Given the diversity inherent in software development, a significant challenge in this domain is selecting the most suitable model. Scholarly work often explores various aspects of software development lifecycles. On the other hand, Application Lifecycle Management (ALM) extends beyond SDLC, incorporating stages that follow the development phase. Integrated ALM seeks to synergize developers and tools throughout the development and operational phases. Research within this realm predominantly investigates ALM's practical applications and impact on the industry. Information Lifecycle Management (ILM) pertains to managing storage systems through comprehensive policies encompassing information creation, receipt, distribution, use, maintenance, and disposition. ILM increasingly draws attention due to rising information security concerns and regulatory requirements. It is frequently embedded within broader lifecycle management frameworks. Product Lifecycle Management (PLM) spans a product's life, from innovation to dismantling. PLM facilitates product complexity management by integrating people, processes, systems, and data, providing valuable feedback on product utilization. This concept is closely aligned with Industry 4.0. Similarly, Service Lifecycle Management (SLM) strives to align service management, communications, and operations to optimize service uptime [14]. Software safety refers to the practice of ensuring that software operates correctly and reliably without causing unintended harm. This is particularly important in industries where software failures, such as automotive, aerospace, healthcare, and industrial control systems, can have catastrophic consequences.

Incorporating software safety into PLM involves several key practices:

- Safety by design: This approach integrates safety considerations into the design phase, ensuring potential risks are identified and mitigated early. By addressing safety from the outset, developers can avoid costly and time-consuming redesigns later in the lifecycle.

- Risk management: Throughout the product lifecycle, it is essential to continually assess and manage risks associated with the software. This involves identifying potential hazards, evaluating their impact, and implementing measures to reduce or eliminate them.

- Compliance with standards: Adhering to industry-specific safety standards, such as ISO 26262 [13] for automotive software, is crucial for ensuring that software meets the necessary safety requirements. These standards provide guidelines for safety-critical software development, including risk assessment, design, testing, and verification processes.

- Continuous monitoring and improvement: Even after the software is deployed, continuous safety monitoring is essential. This involves monitoring performance, identifying new risks, and implementing updates or patches to address any safety issues that may arise. A proactive approach to safety management helps prevent accidents and ensures the software's ongoing safety throughout its lifecycle.

3.3 Challenges in integrating software safety with PLM

Incorporating software safety into PLM is crucial, but it presents several challenges. One primary obstacle is the increasingly complex nature of software systems. As software becomes more sophisticated and interconnected, ensuring its safety becomes more difficult. Integrating software with hardware systems, especially in fields like the IoT, adds further complexity to safety management. Another challenge is the dynamic nature of safety requirements. As technology and user needs change, software safety requirements must also evolve. This calls for a flexible and adaptable approach to PLM, where safety considerations are continuously revisited and updated throughout the product lifecycle. Resource constraints, including time, budget, and expertise, can pose significant challenges. Ensuring software safety often demands additional resources, such as specialized tools, testing procedures, and expert knowledge. Balancing these requirements with project timelines and budgets can be challenging, particularly in fast-paced development environments [17].

To effectively integrate software safety into PLM and overcome challenges, organizations should adopt several best practices:

- Early safety integration: Incorporating safety considerations from the earliest stages of the product lifecycle is crucial for identifying and addressing risks before they escalate. This entails conducting comprehensive safety assessments during the conceptualization and design phases.

- Cross-functional collaboration: Safety assurance is a collective responsibility requiring cohesive efforts across diverse teams, encompassing software developers, product managers, safety engineers, and quality assurance specialists. Maintaining safety as a principal focus throughout the entire lifecycle is contingent upon effective communication and collaboration.

- Agile and iterative development: Incorporating Agile or iterative development methodologies enables the seamless integration and ongoing testing of safety features, facilitating the early identification and resolution of issues. This approach further facilitates the continual enhancement of safety protocols as the software evolves.

- Use of automated tools: Utilizing automated tools for testing, code analysis, and safety assessment can optimize the safety management process and mitigate the potential for human error. These tools promptly identify safety concerns and furnish developers with practical insights.

- Continuous training and education: It is crucial to ensure that all team members know software safety practices and the specific safety requirements of their industry. Continuous training helps keep teams up to date with the latest safety standards and best practices.

Current lifecycle management models only partially address SoS needs. ALM focuses on software, ILM on information flow, SDLC on code development, and SLM on service management. PLM is product-specific and not fully aligned with SoS. Thus, no single model completely meets SoS requirements [14]. The link between PLM and software safety is crucial in developing reliable and secure software products. Safety considerations must be incorporated at every product lifecycle stage to deliver products that meet the highest safety and reliability standards. Establishing practices such as early safety integration, cross-functional collaboration, and continuous monitoring is key. As technology progresses, ongoing research and innovation are necessary to ensure software products' continued safety and success.

4. Machinery safety and PLM

The product management lifecycle serves as a comprehensive framework for guiding product development, launch, and continuous improvement, particularly in manufacturing, construction, and agriculture industries where machinery safety is paramount. Throughout the entire product management lifecycle, machinery safety is of utmost importance. Failing to address safety concerns effectively can lead to accidents, injuries, legal implications, and damage to a company's reputation. It is vital to integrate safety measures in each phase of the lifecycle:

- Ideation stage: During the ideation stage, it is imperative to consider safety as a critical component. This encompasses the identification of potential machinery-related hazards, comprehending the operational environment, and assessing risks to operators and bystanders. Risk assessments evaluate the severity and probability of these hazards with the objective of either eliminating hazards where feasible or integrating safety features to alleviate risks.

- Design and development stage: Safety considerations become more tangible during the design and development phase. Engineers and designers collaborate to ensure compliance with relevant safety standards and regulations, such as those stipulated by the International Organization for Standardization (ISO) or local regulatory entities. This phase incorporates safety features such as emergency stop buttons, guarding, sensors, and fail-safe mechanisms. Prototypes undergo rigorous testing to identify and rectify potential safety issues before mass production [12].

- Production and manufacturing stage: Throughout the production phase, safety considerations extend to the manufacturing machinery and the manufacturing process. Stringent quality control processes are imperative to guarantee adherence to safety specifications for each unit. Moreover, the manufacturing process must be designed to minimize worker risks, incorporating safeguards such as machine guarding, proper ventilation, and ergonomic workstations. Regular inspections and maintenance of production equipment are pivotal in accident prevention and ensuring a safe working environment.

- Launch and market introduction stage: Clear communication of safety information to end users upon product launch is essential. This entails providing comprehensive user manuals, safety warnings, and training materials. Marketing strategies should accentuate the machinery's safety features to foster customer confidence. Vigilant monitoring of customer feedback during this phase is crucial to uncover any safety issues that may have been overlooked during testing.

- Post-market support and continuous improvement: Continuous monitoring is indispensable to ensure sustained safety following the product's market introduction [18]. This encompasses tracking warranty claims, customer complaints, and field reports to identify emergent safety concerns. If safety issues are identified, prompt measures such as product recalls, updates or modifications are imperative. Continuous improvement processes also necessitate updating safety features in line with new technologies, evolving regulations, and customer feedback.

- Product retirement: Machinery's safe disposal or recycling takes precedence after the product lifecycle. This involves ensuring that machinery is decommissioned in a manner that does not pose hazards to workers or the environment. Companies must furnish clear instructions for safe dismantling and disposal procedures and, where viable, promote materials recycling to diminish environmental impact.

4.1 Challenges and best practices in ensuring machinery safety

Ensuring machinery safety throughout the product lifecycle presents several challenges as per IEC 62061 [12]:

- Compliance with regulations: Navigating the complex landscape of safety regulations and standards takes time and effort. Regulations can vary by region and industry and are subject to change. Staying informed and compliant requires ongoing effort and collaboration with regulatory bodies.

- Balancing safety with innovation: Innovating while ensuring safety can be challenging, especially when new technologies are involved. Companies must strike a balance between incorporating cutting-edge features and maintaining a high standard of safety.

- Cost considerations: Implementing comprehensive safety measures can be costly. However, the long-term benefits of preventing accidents, avoiding legal liabilities, and protecting the company's reputation often outweigh the initial investment.

- User behavior: Machinery safety also depends on how users interact with the product. Even the safest machinery can become dangerous if used improperly. Therefore, educating users and providing clear instructions are critical.

4.2 Best practices for ensuring machinery safety

- Early safety integration: It is imperative to embed safety considerations within the product lifecycle from its inception. This proactive approach helps avert safety issues at later stages of the process.
- Routine safety audits: The conduct of regular safety audits during the product lifecycle is critical for the early identification and mitigation of potential risks that may otherwise lead to accidents.
- Technical collaboration: Ensuring safety necessitates collaboration across diverse departments such as design, engineering, manufacturing, and marketing. Involving cross-functional teams facilitates the incorporation of varied perspectives on safety matters.
- Continuous training: Training employees and customers on safety procedures and best practices are pivotal for upholding high safety awareness.

The product management lifecycle offers a systematic approach to product development and management, whereas machinery safety ensures product safety throughout its lifespan. Assimilation of safety considerations at every phase of the product lifecycle is indispensable for producing machinery that aligns with regulatory standards, safeguards users, and upholds a company's standing. By adhering to best practices and tackling safety management challenges, companies can adeptly traverse the complexities of product development and deliver efficacious and secure machinery [12].

5. PLM and functional security

PLM is a strategic approach encompassing a product's lifespan, overseeing its safety and reliability. PLM integrates people, processes, and technologies to manage a product's lifecycle, involving several key stages in the context of software and complex systems [19]. The product development process consists of several key stages. The initial stage involves generating ideas, identifying market needs, and assessing technical feasibility while considering functional security to define security requirements and potential risks early on. The Design and Development phase includes creating detailed designs and developing the product, with functional security integrated into the design process to ensure that security mechanisms are part of the product's architecture [19]. In the Production and Testing stage, the product is manufactured or coded, followed by rigorous testing to validate that the product meets security requirements and functions as intended without vulnerabilities. Once deployed, the product enters the operational phase in operation and maintenance, where it must be maintained and updated to address new security threats and ensure continued safe operation [20]. Lastly, as products reach the end of their useful life in the End-of-Life phase, managing this phase properly is essential to ensure that the product is decommissioned securely, preventing residual security risks.

5.1 The impact of functional security in PLM

Functional security encompasses the measures and practices implemented to ensure a product's proper and safe operation, particularly in environments where malfunctions could lead to significant harm or damage. The RAMS policy, the RAMS management plan, and controlling documents are the overall framework for product development [1]. This is especially critical in industries such as automotive, aerospace, healthcare, and industrial control systems, where security breaches or functional failures can have severe consequences. It is crucial to incorporate functional security into PLM through several key practices:

- Security by design to create inherently secure products, it is crucial to incorporate security considerations into the design phase. This involves identifying potential threats, defining security requirements, and integrating security features into the product from the outset.

- Continual risk assessment is crucial throughout the product lifecycle to identify and address potential security threats. This includes evaluating the impact of security vulnerabilities and implementing measures to reduce or eliminate them.

- Adhering to industry-specific security standards, such as ISO 26262 [13] for automotive functional safety or IEC 62443 [21] for industrial control systems, is crucial for ensuring that products meet necessary security requirements. These standards offer guidelines for developing security-critical products, including risk assessment, design, testing, and verification processes.

- Continuous Monitoring and Adaptation security is an ongoing process that requires continual monitoring and adaptation. Security measures must be updated as new threats emerge and the product environment changes to ensure ongoing protection [22].

5.2 Challenges in incorporating functional security with PLM

Integrating functional security into PLM is crucial but poses several challenges [21]:

- Increasing complexity: Modern products, particularly those involving software and interconnected systems, are becoming more complex. Ensuring comprehensive security throughout the lifecycle is challenging. For example, integrating IoT devices with cloud services requires robust security measures to safeguard data and functionality across various platforms.

- Evolving threats: Security threats are constantly evolving, making it difficult to anticipate all potential risks during the initial stages of the product lifecycle. This demands a dynamic security approach capable of adapting to emerging threats.

- Resource constraints: Ensuring functional security throughout the product lifecycle often demands significant resources, including time, expertise, and financial investment. Balancing these needs with project constraints can be challenging, especially in industries with critical time-to-market requirements.

- Interdependencies: Different components and subsystems within complex systems are often interdependent. Ensuring the functional security of the entire system necessitates careful coordination and management of these interdependencies, as a failure in one component can jeopardize the entire system's security.

5.3 Best practices for ensuring functional security in PLM

Organizations aiming to integrate functional security into PLM seamlessly should adhere to several key best practices:

- Early integration of security measures: The inclusion of functional security considerations should commence at the inception of the product lifecycle. This involves the identification of security requirements and potential risks during the conceptual and design phases, thereby facilitating the incorporation of security measures from the outset, consequently reducing the probability of vulnerabilities at later stages of the lifecycle [21].

- Collaboration: Security is a communal responsibility necessitating collaboration across diverse teams encompassing product management, design, engineering, and quality assurance. By nurturing a culture of heightened security awareness and collaboration, organizations can ensure the prioritization of security throughout the lifecycle.

- Agile and iterative development: Organizations can consistently integrate and test security features by embracing Agile or iterative development practices. This enables early detection and rectification of security issues and supports the continual enhancement of security measures as the product matures.

- Use of automated tools: Deploying automated tools for security testing, code analysis, and vulnerability assessment can streamline the security management process by swiftly identifying potential security issues and providing developers with actionable insights.

- Continuous training: It is essential to sustain comprehensive knowledge of functional security practices. Regular training and educational programs keep teams abreast of the latest security standards and best practices, thus ensuring the sustained prioritization of security throughout the product lifecycle [17].

Notably, Product Lifecycle Management and functional security are fundamentally interconnected in developing secure and dependable products. As products evolve to become increasingly intricate and interconnected, the imperative of integrating functional security into each stage of the product lifecycle cannot be overstated. Organizations can effectively manage functional security and deliver products meeting the highest safety and reliability standards by embracing best practices such as early security integration, cross-functional collaboration, and persistent monitoring. As technology advances, the connection between PLM and functional security will only grow significantly, requiring continuous innovation and adaptation to ensure contemporary products' perpetual safety and success.

6. Improving product lifecycle RAMS requirements using Industry 4.0

The RAMS principle is pivotal in developing and managing products, particularly in industries where high reliability and safety are paramount, such as aerospace, automotive, and industrial automation. With the advent of Industry 4.0, a revolutionary paradigm integrating advanced technologies like the IoT, Artificial Intelligence (AI), big data analytics, and smart manufacturing, there is an unprecedented opportunity to enhance RAMS requirements throughout the entire product lifecycle. This section delves into how Industry 4.0 technologies can enrich RAMS requirements, ultimately ensuring more reliable, available, maintainable, and safer products [3]. Historically, RAMS requirements are established during the design phase and monitored throughout the product lifecycle. However, Industry 4.0 technologies present new methods to enhance these requirements dynamically, leading to more robust and adaptive systems, as shown in **Figure 4**.

The digital thread involves digitizing and tracing a product throughout its lifecycle, connecting digital twins' capabilities, designs, performance data, product data, supply chain data, and software. This integration allows for identifying issues, incorporating feedback, and isolating causes of failures, all of which are crucial for product development and maintenance [10].

6.1 Leveraging Industry 4.0 for enhanced RAMS

Industry 4.0 introduces several key technologies that can significantly improve RAMS requirements across the product lifecycle:

- IoT: The wireless-based system connects devices and systems, facilitating real-time data collection and communication. In the context of RAMS, IoT can be utilized to monitor the condition of products in operation continuously.

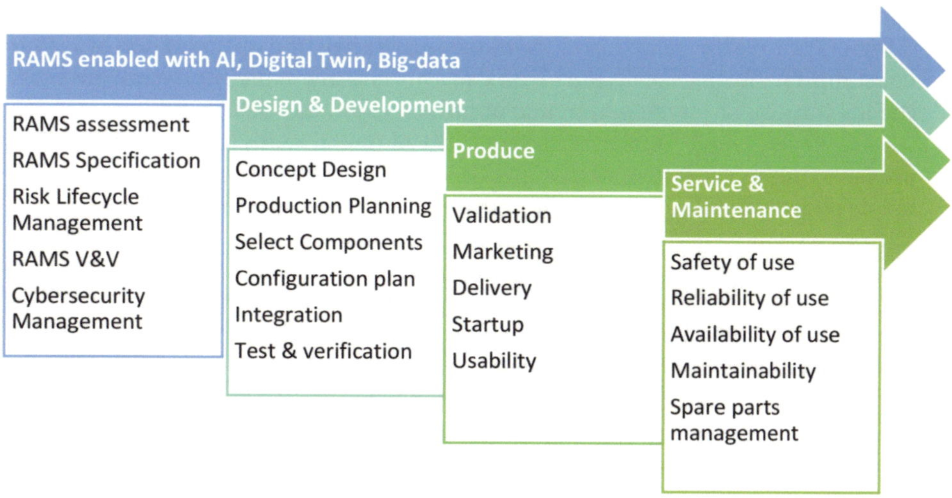

Figure 4.
RAMS relation to the major stages.

Embedded sensors can gather data on performance, environmental conditions, and wear and tear, offering valuable insights into reliability and availability. This real-time data enables predictive maintenance, where potential failures are identified and addressed before they occur, ultimately enhancing reliability and maintainability [2].

- Big data analytics: The significant amount of data generated by IoT devices can be analyzed using big data analytics. This empowers companies to discover patterns and trends that impact RAMS by processing and analyzing the data. For example, big data analytics can uncover the root causes of failures, optimize maintenance schedules, and predict future reliability issues. This data-driven approach enables more informed decision-making and proactive management of RAMS requirements.

- Artificial Intelligence (AI) and Machine Learning (ML): AI and ML algorithms have the potential to enhance RAMS by automating the analysis of complex datasets and making predictions based on historical and real-time data. These algorithms can create predictive models to estimate the remaining useful life of components, identify potential safety risks, and optimize system availability. Additionally, AI can be utilized in the design phase to simulate various scenarios and optimize RAMS requirements before constructing the product [23].

- Digital twins: A digital twin serves as a virtual replica of a physical product or system, providing an up-to-date representation of its real-world counterpart. Companies leverage digital twins to simulate and analyze product performance under various conditions, ultimately enhancing reliability and safety assessments. Digital twins also play a key role in evaluating maintenance strategies, leading to reduced downtime and improved availability [24, 25]. This continuous feedback loop between the physical product and its digital twin allows for dynamic adjustments to RAMS requirements as the product operates in the real world. The sustainability of a product throughout its life cycle is heavily influenced by its design. The mindful selection of sustainable materials and optimization of resource consumption during the design phase can significantly reduce its overall environmental impact. Embracing a circular approach to product design, which emphasizes reusing or recycling, coupled with the ability of digital twins to assess end-of-life scenarios, holds the potential to improve sustainability greatly. This approach drives the development of innovative and environmentally responsible products [26].

- Smart manufacturing and automation: Industry 4.0 technologies, including robotics, automation, and smart manufacturing, play a crucial role in enhancing the production process by ensuring consistent quality and minimizing the likelihood of defects that could impact reliability, availability, maintainability, and safety. Automated systems are capable of monitoring production processes in real-time, promptly detecting and rectifying any deviations from the desired specifications. This improves product reliability and safety while ensuring that items are manufactured with easy maintenance. Creating complex systems requires collaboration among large teams of engineers and stakeholders. Synchronizing efforts through connected work environments leads to better results and facilitates information exchange. A digital environment is an integrated ecosystem of data interaction that supports a digital thread, utilizing digital models and tools for design, analysis, simulation, manufacturing, and sustainment activities across the product lifecycle [10, 26].

- Augmented Reality (AR): AR technologies can enhance maintainability by providing technicians with immersive training and real-time guidance during maintenance tasks. For example, AR can overlay digital information onto physical equipment, enabling technicians to promptly identify issues and follow maintenance procedures. This helps reduce the likelihood of human error, enhances safety, and ensures efficient product maintenance.

6.2 Challenges and considerations

Although Industry 4.0 technologies offer significant benefits for enhancing RAMS requirements, several challenges need to be taken into account:

- Data security and privacy: With the increased connectivity and data exchange inherent in Industry 4.0, products and systems are more vulnerable to cybersecurity threats. It is crucial to ensure that data is securely transmitted, stored, and processed to maintain the integrity of RAMS requirements.

- Integration and interoperability: Seamlessly integrating different systems and platforms is essential for effectively implementing Industry 4.0 technologies to enhance RAMS. However, this also presents a significant challenge in preventing the introduction of new vulnerabilities or complexities.

- Cost and resource allocation: Deploying Industry 4.0 technologies requires substantial infrastructure, software, and training investments. Companies need to carefully evaluate the costs concerning the potential benefits for RAMS requirements and determine the most efficient allocation of resources.

- Change management: Adopting new technologies and processes can be particularly challenging for organizations operating in industries with established practices and regulations. Implementing effective change management strategies ensures that the transition to Industry 4.0 enhances, rather than disrupts, RAMS requirements.

Industry 4.0 signifies a profound product design, manufacturing, and management transformation across their lifecycle. Organizations can significantly enhance RAMS requirements by utilizing technologies such as IoT, big data analytics, AI, digital twins, and smart manufacturing, thereby yielding more reliable, available, maintainable, and safer products. However, realizing these benefits necessitates meticulous consideration of data security, integration, cost, and change management challenges. As Industry 4.0 advances, its capacity to elevate RAMS requirements will continue to expand, presenting novel prospects for innovation and excellence in product lifecycle management.

7. Summary

In summation, PLM and RAMS constitute pivotal frameworks that substantially augment the management and performance of products throughout their lifecycle.

PLM is concerned with the exhaustive management of a product from its conceptualization through its design, manufacturing, service, and disposal, thereby facilitating enhanced collaboration, innovation, and efficiency. Conversely, RAMS underscores the imperative of ensuring that products are reliable, available, maintainable, and safe, which is indispensable for satisfying customer expectations and complying with regulatory requisites.

Integrating PLM and RAMS frameworks can streamline processes, enhance product quality, and improve customer satisfaction. Utilizing data and insights from both frameworks enables organizations to make informed decisions, optimizing product performance and lifecycle management. As industries adapt to emerging technologies and changing market demands, the synergistic relationship between PLM and RAMS will become increasingly critical for achieving sustainable growth and maintaining a competitive advantage.

Acronyms and abbreviations

SDLC	software development lifecycle management
AI	artificial intelligence
ALM	application lifecycle management
AR	augmented reality
ERP	enterprise resource planning
FMEA	failure modes and effects analysis
FTA	fault tree analysis
HAZOP	hazard and operability study
ILM	information lifecycle management
IoT	internet of things
MBSA	model-based safety assessment
MBSE	model-based systems engineering
ML	machine learning
MTBF	mean time between failures
MTTR	mean time to repair
PLM	product lifecycle management
RAMS	reliability, availability, maintainability, and safety
SLM	service lifecycle management

Author details

Kourosh Parsa* and Mohsen Naderpour
University of Technology Sydney, Sydney, NSW, Australia

*Address all correspondence to: koupa@ieee.org

References

[1] Lundteigen MA, Rausand M, Utne IB. Integrating RAMS engineering and management with the safety life cycle of IEC 61508. Reliability Engineering & System Safety. 2009;**94**(12):1894-1903

[2] Chen D. An overview method of viable dependability for RAMS in life cycle. In: 2018 Annual Reliability and Maintainability Symposium (RAMS). IEEE; 2018. pp. 1-7

[3] Elangovan U. Product Lifecycle Management (Plm): A Digital Journey Using Industrial Internet of Things (Iiot). CRC Press; 2020

[4] IEC 61508, Functional safety of electrical/electronic/programmable electronic safety related systems, I. E. Commission, 2016

[5] IEC 61511, Functional safety—Safety instrumented systems for the process industry sector, I. E. Commission, 2016

[6] Lalonde C, Boiral O. Managing risks through ISO 31000: A critical analysis. Risk Management. 2012;**14**:272-300

[7] dos Santos KCP, de Freitas Rocha Loures E, Junior OC, Santos EAP. Product lifecycle management maturity models in Industry 4.0. In: Product Lifecycle Management to Support Industry 4.0: 15th IFIP WG 5.1 International Conference, PLM, Proceedings 15, 2018. Turin, Italy: Springer; 2018. pp. 659-669

[8] Unver HO. An ISA-95-based manufacturing intelligence system in support of lean initiatives. The International Journal of Advanced Manufacturing Technology. 2013;**65**:853-866

[9] Shivankar SD, Deivanathan R. Product design change propagation in automotive supply chain considering product life cycle. CIRP Journal of Manufacturing Science and Technology. 2021;**35**:390-399

[10] Beshears R, Long J. Digital engineering transformation for reliability/maintainability. In: 2024 Annual Reliability and Maintainability Symposium (RAMS). IEEE; 2024. pp. 1-5

[11] Bukowski JV, Goble WM. Predictive analytics: Assessing failure rate accuracy & failure mode completeness. In: Reliability and Maintainability Symposium (RAMS), 2013 Proceedings-Annual. IEEE; 2013. pp. 1-7

[12] I. E. Commission. IEC62061. Safety of Machinery—Functional Safety of Safety-Related Electrical, Electronic, and Programmable Electronic Control Systems. Switzerland: IEC Geneva; 2005

[13] I. S. Organization. ISO 26262—Functional Safety—Automotive. ISO. 2018

[14] Kozma D, Varga P, Larrinaga F. System of systems lifecycle management—A new concept based on process engineering methodologies. Applied Sciences. 2021;**11**(8):3386

[15] INCOSE. INCOSE systems engineering handbook. John Wiley & Sons; 2023

[16] Leveson N. Safety III: A systems approach to safety and resilience. MIT Engineering Systems Lab. 2020;**16**:2021

[17] Leveson N. Engineering a Safer World: Systems Thinking Applied to Safety. MIT press; 2011

[18] Kafka P. The automotive standard ISO 26262, the innovative driver for enhanced safety assessment & technology for motor cars. Procedia Engineering. 2012;**45**:2-10

[19] I. I. I. 15288. Systems and Software Engineering—System Life Cycle Processes. ISO/IEC/IEEE; 2023

[20] Leander B, Čaušević A, Hansson H. Applicability of the IEC 62443 standard in Industry 4.0/IIoT. In: Proceedings of the 14th International Conference on Availability, Reliability and Security. 2019. pp. 1-8

[21] I. E. C. (IEC). 62443 Security for Industrial Automation and Control Systems. 2023

[22] I. E. Commission. IEC 62682 Management of Alarm Systems for the Process Industries IEC. Switzerland: Geneva; 2014. pp. 24-26

[23] Urgese G, Azzoni P, van Deventer J, Delsing J, Macii E. An engineering process model for managing a digitalised life-cycle of products in the industry 4.0. In: NOMS 2020-2020 IEEE/IFIP Network Operations and Management Symposium. IEEE; 2020. pp. 1-6

[24] Uhlemann TH-J, Lehmann C, Steinhilper R. The digital twin: Realizing the cyber-physical production system for industry 4.0. Procedia CIRP. 2017;**61**:335-340

[25] Vogel-Heuser B, Ocker F, Weiß I, Mieth R, Mann F. Potential for combining semantics and data analysis in the context of digital twins. Philosophical Transactions of the Royal Society A. 2021;**379**(2207):20200368

[26] Deuse J, Wagstyl D, Moreno VH, Polikarpov M, Wöstmann R, Hoffmann F. Green digital twins in the product life cycle. In: Schriftenreihe der Wissenschaftlichen Gesellschaft für Arbeits-und Betriebsorganisation (WGAB) e. V. 2023. pp. 167-186

Chapter 4

Enhancing Pipeline Reliability: A Structural Integrity Management Approach Using Minimal Cut Set Method and Importance Measures

Marco Antônio Sabará,
Jose Antônio da Cunha Ponciano Gomes and
Alysson Helton Santos Bueno

Abstract

Gas pipelines are fundamental structures for transporting energy resources. Their integrity is constantly threatened by failures caused by potential punctures or ruptures, leading to gas releases, which can have significant consequences for the installation, people, and the environment. Various methodologies have been proposed to improve Pipeline Structural Integrity Management (PSIM) processes. In this work, a model for estimating the probability of gas release failure is proposed using a Quantitative Fault Tree Analysis (QFTA) approach. The Minimum Cut Set (MCS) technique is applied along with the assessment of Importance Measures (IM) to provide an accurate estimation of the failure rate (λ) and the identification of the most critical basic events. This information can be used to support actions in Risk-Based Inspection (RBI) and Reliability-Centered Maintenance (RCM) to eliminate, control, or mitigate potential risks. The model was validated by comparing the results obtained through Monte Carlo Simulation with data from official databases of pipeline incidents/accidents and similar models published in the literature. The proposed model proved capable of accurately estimating the failure rate (λ), with results closely matching the database values and more convergent than those achieved by reference models in the literature. The study also provides guidelines for the correct and effective application of the model in pipeline PSIM routines.

Keywords: structural integrity management, risk management, minimal cut set method, Monte Carlo simulation, reliability, corrosion

1. Introduction

Gas pipelines are critical infrastructures for the transportation of energy resources, but their integrity is constantly challenged by various threats, such as corrosion, material/weld/equipment failure, and Third-Party Damage [1, 2]. Improving the reliability of gas pipelines is essential to avoid costly failures and environmental damage, and ensure safe operations [3, 4].

This paper proposes a novel approach to gas pipeline integrity management using the minimal cut set (MCS) method and importance measures. The MCS identifies the most critical failure scenarios, while importance measures quantify each component's (or basic events) contribution to system reliability [5, 6]. By focusing on the most influential factors, this approach optimizes resource allocation for inspection and maintenance, thereby enhancing the overall reliability of gas pipelines [2, 4, 7].

The study integrates reliability engineering principles with the design, operation, and maintenance of gas pipelines to develop a comprehensive, reliability-centered framework [1, 3, 8]. This includes the following:

- Identifying critical failure modes through MCS analysis
- Prioritizing components based on importance measures
- Optimizing inspection intervals and maintenance strategies
- Validating the approach through case studies and industry data

The proposed methodology provides gas pipeline operators with a systematic, data-driven tool to manage integrity and reduce failure risks [2, 4, 9]. By targeting the most influential factors, it allows efficient resource allocation while maintaining high reliability standards [1, 3, 8].

This research contributes to the growing body of knowledge on gas pipeline integrity management, offering a practical solution to improve the safety and sustainability of energy transportation infrastructure [2, 4, 5, 10]. The integration of MCS and importance measures represents a significant advancement in the field, with potential applications in other critical systems [10].

Despite rigorous design standards and meticulous maintenance protocols, the inherent risks associated with gas pipeline operations demand robust risk assessment methodologies [3, 5]. Traditional approaches provide valuable insights [7, 10], but as pipeline networks expand and operational conditions become more complex, there is an increasing need for advanced quantitative risk assessment models [1, 2, 4]. In this context, a comprehensive model has been developed to determine the global failure probability of pipeline systems, leveraging the strengths of Quantitative Fault Tree Analysis (QFTA).

Quantitative Fault Tree Analysis (QFTA) is a systematic and deductive method that offers a granular understanding of potential failure mechanisms within complex systems [7, 10]. By breaking down the system into its constituent components and mapping out the logical relationships between potential failures, QFTA allows for a detailed probabilistic analysis of system reliability. In the proposed model, the application of QFTA is extended to the domain of pipeline systems, incorporating both operational and environmental variables to reliably estimate failure probabilities [7]. This model not only enhances the predictive accuracy of failure scenarios but also provides a valuable tool for optimizing maintenance strategies and improving overall system resilience. It achieves this by understanding the events that contribute to various failure mechanisms in pipelines and establishing their criticalities through the use of importance measures [1, 2, 4, 7, 10].

2. A brief review based on the construction of a fault tree analysis

Failures in gas pipelines can occur due to various mechanisms and with different failure rates [11]. The probability of failure is significantly influenced by design, construction, operation, maintenance techniques, and the environmental conditions surrounding the gas pipeline [11, 12].

Considering gas release as the potentially most severe consequence of these failures [4–6], it is possible to identify puncture and rupture as the failure modes that most expose gas pipelines to risk and allow it to propagate through the system, leading to consequences for human life, society, the environment, and the economy [11, 13, 14].

Fault Tree Analysis (FTA) is a practical and logical method for evaluating the complexity of interactions among the various events that can lead to these failure modes [15]. This technique also allows for the identification of the most critical events related to these failure modes and their impact on the risk of gas release through a quantitative approach, thus supporting the development of targeted mitigation strategies [11, 16, 17].

Based on the literature [11–14, 16, 17], a pipeline failure tree is proposed (**Figure 1**). Several factors can be identified as contributing to pipeline failure, primarily: third-party activities (accidental, incidental, malicious, sabotage, act of war, etc.), corrosion (external or internal, with and without the action of external mechanical effects), hydrogen damages (hydrogen embrittlement and cathodic disbondment), discontinuities, operational defects (system failures and human errors, including inadequate maintenance), and natural hazards (earthquake, flood, ground subsidence, ground movement, etc.).

This information allows for the construction of a Fault Tree Analysis to better understand the problem (**Figure 1**).

It is possible to observe in **Figure 1** that the failure modes Puncture and Rupture are correlated by a logical OR gate. This indicates that the occurrence of just one of these failure modes is sufficient for the top event to occur. Therefore, a gas release occurs if there is either a puncture or a rupture in the gas pipeline; or, alternatively, if both conditions occur simultaneously.

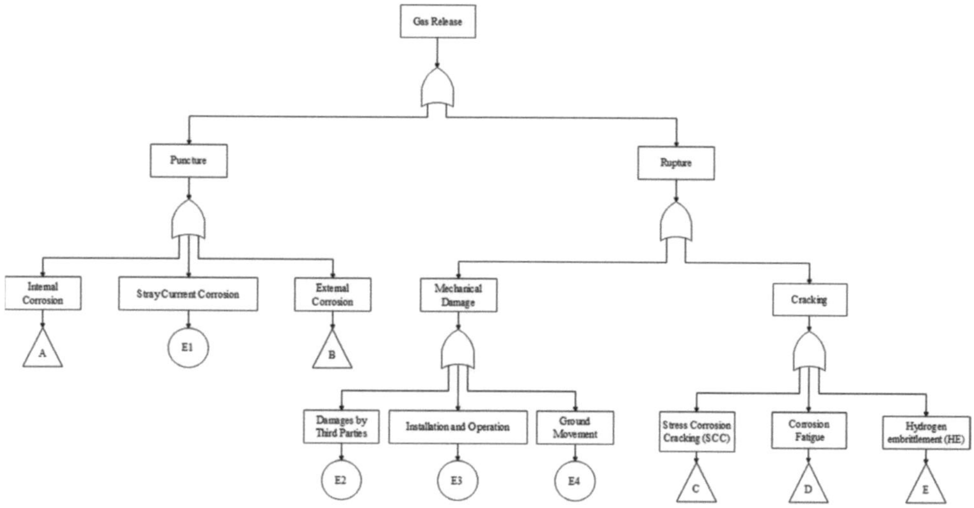

Figure 1.
Gas release in gas pipelines FTA (overview).

The Puncture failure mode (second level, left side of the FTA in **Figure 1**) is related to the failure mechanisms of Internal Corrosion, Stray Current Corrosion, and External Corrosion, also through an OR gate. This means that the occurrence of any one of these mechanisms is sufficient for the failure mode to be activated.

The Rupture failure mode (second level, right side of the FTA in **Figure 1**) is related to the failure mechanisms of Mechanical Damage and Cracking, also through an OR gate, meaning that the occurrence of either mechanism is sufficient for the failure mode to be activated.

The Mechanical Damage failure mechanism (third level, bottom center of **Figure 1**) is triggered by Third-Party Damage, operational and installation defects, or ground movement events (OR gate). The Cracking failure mechanism, on the other hand, is activated if Stress Corrosion Cracking (SCC), Corrosion Fatigue, or Hydrogen Embrittlement (HE) during gas pipeline operation occur (OR gate).

Stray Current Corrosion (E1), Third-Party Damage (E2), Installation and Operational defects (E3), and Ground Movement (E4) are basic events. According to IEC 61025:2006 [15], these events are at the lowest level of resolution in the fault tree and do not need to be further developed or decomposed. They are primary events for which the probability of occurrence is assumed to be known or can be estimated through some reliable methodology [13–15].

Stray Current Corrosion (Basic Event E1 in **Figure 1**) is a critical phenomenon in pipeline integrity degradation, particularly in gas transmission systems [9–11]. This type of corrosion occurs when unintended electrical currents, typically originating from external sources such as electric railways, improperly grounded electrical systems, or cathodic protection systems (CPS) of adjacent structures, pass through the pipeline material. These currents induce electrochemical reactions that accelerate the corrosion process at specific locations on the pipeline [18, 19].

The localized nature of stray current corrosion makes it particularly insidious. Unlike uniform corrosion, which affects large areas of the pipeline uniformly, stray current corrosion can create deep pits or localized areas of metal loss. These defects significantly reduce the structural integrity of the pipeline, making it susceptible to puncture or rupture under operational pressures [20, 21].

Stray current corrosion is a potent failure mechanism contributing to gas release in gas pipelines [9, 10]. It initiates corrosion pits that penetrate the pipeline wall, creating small holes or fissures through which gas can escape. These leakage points can be challenging to detect until significant gas loss occurs, and once initiated, corrosion pits can propagate rapidly, exacerbating the severity of the leakage. The presence of high-pressure gas within the pipeline accelerates the expansion of these pits into larger cracks or ruptures, amplifying the risk to public safety, the environment, and the operational continuity of the gas transmission system [19, 21, 22].

Effective mitigation of stray current corrosion involves several strategies. This includes implementing cathodic protection systems (CPS), such as sacrificial anode or impressed current systems, to neutralize stray currents and protect the pipeline material. Electrical isolation, through the installation of insulating joints or coatings, can prevent the entry and exit of stray currents from the pipeline. Regular monitoring of pipeline potentials and environmental conditions, combined with proactive maintenance and repair of identified corrosion sites, is also crucial [18, 21, 22].

Third-Party Damage (Basic Event E2 in **Figure 1**) represents a significant concern in the context of pipeline integrity [23, 24]. This type of damage occurs due to activities or events external to the gas pipeline infrastructure. Examples include excavation activities, construction projects, agricultural operations, or even deliberate acts of vandalism [25, 26].

Third-Party Damage typically manifests through several mechanisms. Firstly, mechanical impact from excavation equipment, construction machinery, or agricultural tools can directly affect the pipeline, causing dents, scratches, or punctures. These mechanical impacts weaken the pipeline structure and may compromise its ability to withstand operational pressures [24]. Secondly, operational errors during excavation or construction activities, such as inaccurate digging procedures or failure to follow safety guidelines, can result in unintended contact with the pipeline. These errors can cause immediate damage or create conditions that contribute to future pipeline failures [24, 25]. Lastly, intentional acts of vandalism, sabotage, or unauthorized interference with the pipeline pose significant risks. These malicious actions can range from deliberate physical damage to tampering with protective coatings or cathodic protection systems, thereby accelerating corrosion and weakening the pipeline over time [24, 26].

Third-Party Damage poses substantial risks to gas pipeline integrity and can directly contribute to gas leak failures [25, 26]. Physical damage from third-party activities can initiate structural defects such as cracks, fractures, or punctures in the pipeline wall, providing pathways for gas release into the surrounding environment. Damage caused by third-party activities may not be immediately detected, especially if the pipeline is buried underground or located in remote areas. Delayed detection prolongs the exposure to risk, potentially resulting in more extensive damage and greater environmental impact. Over time, repeated instances of Third-Party Damage can weaken the overall structural integrity of the pipeline, increasing the likelihood of catastrophic failures that endanger public safety, environmental sustainability, and gas supply continuity [24].

Effective mitigation of Third-Party Damage involves proactive measures to minimize risk and enhance pipeline resilience [24]. Stakeholders involved in excavation, construction, and maintenance activities should receive training on pipeline awareness and safety protocols to prevent accidental damage. Adherence to regulatory guidelines and industry standards regarding pipeline protection, marking, and monitoring is essential to mitigate the risk of Third-Party Damage. Utilizing advanced technologies such as pipeline monitoring systems, geospatial mapping, and remote sensing can improve the detection and response to third-party activities near pipelines [23–27].

Mechanical damage occurring during the installation and operation phases of pipelines (Basic Event E3 in **Figure 1**) also poses threats to the structural and operational integrity of these systems [25]. During the installation phase, errors in the welding process can result in defective or brittle joints, which can become vulnerable points over time. Additionally, manufacturing defects such as weld inclusions or undetected cracks during quality control can compromise the pipe's strength, increasing the risk of premature failures [24]. During operation, failures in control and monitoring systems can lead to undetected adverse operational conditions, such as abnormal pressures or temperature variations in the pipeline, which can accelerate corrosive or mechanical processes. Moreover, improper use or lack of maintenance of cathodic protection systems, designed to protect the pipeline from external corrosion, can allow localized corrosion points to form, eventually leading to structural failures [26]. Furthermore, inadequate maintenance activities, including deficient inspections or extended intervals between preventive maintenance cycles, may result in the non-detection of incipient defects or worsening of existing damages. Inadequate training for operators and maintenance teams can also contribute to errors during pipeline operation and maintenance, thereby increasing the potential for failures [28].

Mitigating damage during the installation and operation of pipelines requires a holistic approach that spans from improving manufacturing and installation processes to implementing robust maintenance and training programs. Adoption of advanced monitoring technologies and strict compliance with standards and regulations are essential to ensure safety, operational efficiency, and environmental sustainability of gas transmission systems [24, 26, 28].

Ground movement (Basic Event E4 in **Figure 1**) represents a serious threat to the structural integrity of buried gas pipelines [28, 29]. This phenomenon can occur due to a variety of geological and environmental factors, including landslides, subsidence, earthquakes, and even human activities such as excavation and mining. The impact of ground movement on buried pipelines is significant and multifaceted, affecting both the physical integrity of the pipeline and its operational capability [30, 31].

When ground movement occurs, pipelines are subjected to stresses and deformations for which they may not be designed. Mass ground movements, such as landslides, can induce excessive longitudinal and transverse stresses in the pipelines, leading to plastic deformations. These deformations can result in ovalization of the pipe, excessive bending, and, in extreme cases, fractures or catastrophic ruptures [31].

Repeated ground movement, such as that caused by frequent earthquakes or freeze-thaw cycles, can induce fatigue in the pipeline material. Fatigue can manifest in the form of microcracks that, over time and under continuous operating conditions, can grow and coalesce, resulting in structural failures. Additionally, repetitive interaction between the pipeline and the soil can lead to wear and degradation of the pipeline's protective coating, exposing it to corrosive agents [28, 29].

Welds and joints in pipelines are particularly vulnerable to ground movement. Excessive stresses and deformations can compromise the integrity of the welds, resulting in cracks or discontinuities that serve as initiation points for failure. A failure in a weld can propagate rapidly, leading to leaks or complete ruptures of the gas pipeline [30].

Furthermore, ground movements can damage the pipeline's protective coating, exposing the underlying metal to a corrosive environment. Exposure to soil and water can accelerate corrosion processes, especially in areas where cathodic protection may have been compromised due to ground movement. Localized corrosion can evolve into pitting, resulting in structural weakness points along the pipeline [28, 31].

To mitigate the impacts of ground movement on the integrity of buried pipelines, it is crucial to implement a series of preventive and corrective measures. Geotechnical monitoring, for example, allows for real-time detection of ground movements and assessment of their potential implications for pipelines. Developing pipelines with materials and construction techniques that allow for greater flexibility and deformation resistance is another important strategy. Additionally, the use of advanced coatings and robust cathodic protection can prevent corrosion, while regular inspections and preventive maintenance ensure that minor damages are corrected before they evolve into major failures [28–31].

The mechanisms of Internal Corrosion, External Corrosion, Stress Corrosion Cracking (SCC), Corrosion Fatigue, and Hydrogen Embrittlement (HE), according to the same IEC 61025:2006 standard [15], are intermediate events. Therefore, they need to be evaluated in their developments until the basic events related to them can be identified [13].

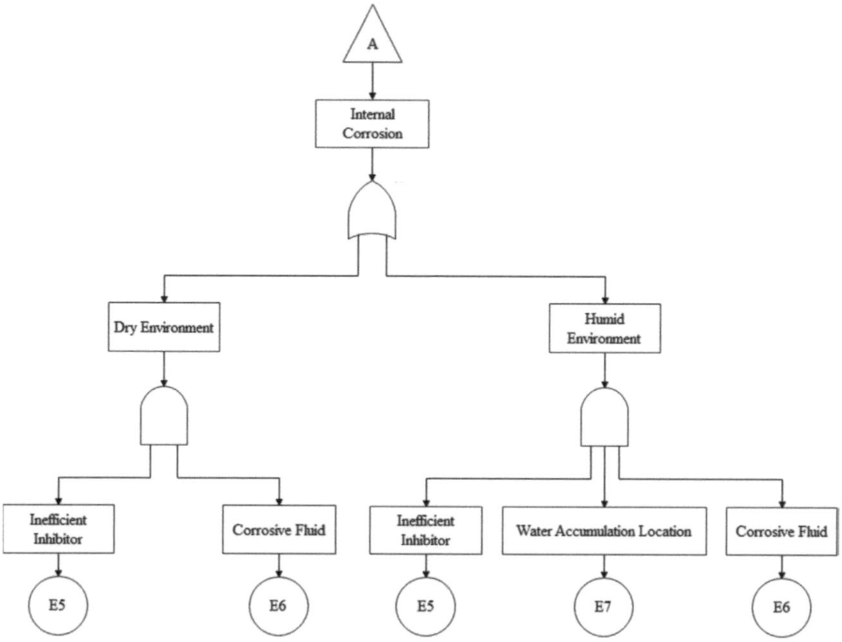

Figure 2.
Failure mechanism—internal corrosion.

Figure 2 proposes the development of the fault tree for the internal corrosion mechanism, based on the literature [29, 32–35].

The mechanism of internal corrosion in gas pipelines manifests distinctly in dry and humid environments (refer to the use of the OR gate in the FTA presented in **Figure 2**). The basic events (bottom line of **Figure 2**) involved in this mechanism include the ineffectiveness of the inhibitor (Basic Event E5), the corrosiveness of the fluid (Basic Event E6), and the existence of a water accumulation site (Basic Event E7) for humid environments. The use of the AND gate indicates that the simultaneous action of more than one basic event is necessary to activate the failure mechanism. The transfer symbol A (at the top center of **Figure 2**) indicates where this branch fits into the FTA presented in **Figure 1** (see the left central region of **Figure 1**).

In gas pipelines, internal corrosion primarily occurs due to the interaction of metallic materials with transported fluids, influenced by a variety of factors [29, 32, 35]. Internal corrosion can be attributed to contaminants present in the transported fluid, which may include acetic acid, gases (CO_2 and H_2S), water, dissolved salts, and other corrosive chemical compounds. The corrosiveness of the fluid is a critical factor, as it determines the rate and type of corrosion that occurs [24]. In environments with humid gases, the presence of water or water vapor is especially problematic, as it can create accumulation points in low areas of the pipeline, where localized corrosion may intensify [32, 33, 35].

The efficacy of corrosion inhibitors, chemical substances designed to slow down or prevent corrosion, also plays a crucial role in mitigating this problem. However, inhibitor failure can occur for various reasons, such as inadequate concentrations, degradation over time, or complex chemical reactions with contaminants present in the fluid [32, 34, 35].

Internal corrosion can manifest broadly, affecting extensive areas of the pipeline's internal surface, or locally, where specific areas suffer greater degradation. Localized corrosion, such as pitting corrosion, can be particularly concerning as it can lead to perforation failures in localized areas, even in materials generally resistant to general corrosion [24, 29, 33].

Figure 3 presents the development of the fault tree for the external corrosion mechanism, based on the literature [36–46].

The mechanism of external corrosion in gas pipelines depends on the simultaneous occurrence of the Basic Soil Corrosivity event (E8) and the ineffectiveness of protective measures, namely: coating failure and cathodic protection failure (E9), as indicated by logical connectives in **Figure 3**. In turn, coating failures can occur due to at least one of the following factors: mechanical damage, natural wear (E10), or cathodic disbondment (E11), as depicted by the logical OR connective at the bottom of **Figure 3**. Symbol B (top left part of **Figure 3**) indicates where this branch intersects the Fault Tree Analysis (FTA) presented in **Figure 1** (see left, central region of **Figure 1**).

External corrosion in gas pipelines represents a significant challenge, potentially compromising the structural integrity and operational safety of these critical infrastructures [36, 40]. This type of failure arises from the interaction of pipeline materials with the external environment, leading to metal degradation over time. Additionally, microorganisms present in soil or adjacent water bodies can accelerate

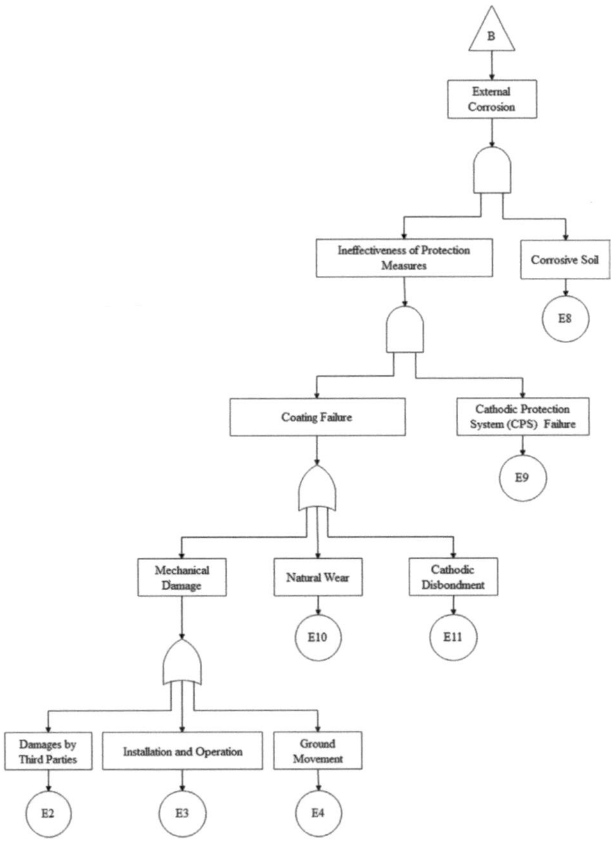

Figure 3.
Failure mechanism—external corrosion.

corrosion processes through Microbiologically Influenced Corrosion (MIC), producing corrosive metabolites that harm metal [39, 43, 44].

Various processes contribute to external corrosion formation, ranging from uniform corrosion affecting the entire metal surface to the development of pits and crevices, which pose challenges in detection and control. Another common issue is undercoating corrosion, occurring when failures or damages in the pipeline's protective coating expose metal to corrosive environments [39, 42–44].

For external corrosion to occur in gas pipelines, a combination of protective measure failures and exposure to a corrosive environment, such as areas with corrosive soil, is necessary. The effectiveness of protective measures is crucial in preventing pipeline material degradation over time [40, 42].

One of the primary protective elements is protective coatings play a critical role in corrosion prevention. However, these coatings can fail due to various reasons. Mechanical damage is a common cause, occurring during installation, handling, or due to soil movements. Natural aging of the coating can also diminish its effectiveness over time, compromising its barrier properties against corrosive environments [40, 45, 46].

Furthermore, Cathodic Protection Systems (CPS) is widely used to mitigate corrosion in buried metal structures. According to NBR 15589 standards, CPS is deemed ineffective when it fails to maintain the electrical potential of the metallic structure within specified protection limits. Specifically, for steel structures, the standard criterion for adequate protection is a potential of −850 mV until −1100 mV versus a Cu/ $CuSO_4$ reference electrode. This may result from inadequate monitoring, faults in sacrificial anodes, or rectifiers supplying the system [36, 38].

Another failure mode is cathodic disbondment, inadequate adhesion, or application process faults. This disbondment creates exposed metal areas vulnerable to localized corrosion especially when combined with moisture and corrosive ions in the surrounding environment [38, 42, 45, 46]. Additionally, repair programs for damaged coatings and replacement of corroded sections are crucial for maintaining the structural integrity of pipelines over time [38, 40, 43, 46].

Continuous monitoring of environmental conditions surrounding pipelines is also essential to identify and address changes that may affect corrosion rates. Selecting appropriate materials with corrosion-resistant properties compatible with specific operating environments is an additional preventive measure that can be implemented during pipeline design. Ensuring the integrity and effectiveness of protective measures is essential in preventing external corrosion in gas pipelines. This involves not only the proper implementation of cathodic protection systems and protective coatings but also regular maintenance, inspection, and monitoring of environmental conditions around pipelines. Additionally, investing in advanced inspection technologies and preventive maintenance programs helps identify early signs of protective measure failures, thereby ensuring the safe and durable operation of pipelines [39, 43, 44].

Figure 4 presents the development of the fault tree for the stress corrosion cracking mechanism, based on the literature [25, 47–56].

The mechanism of Stress Corrosion Cracking (SCC) in gas pipelines relies on the simultaneous action of mechanical deformation and corrosion processes, as indicated by the AND gate in **Figure 3**. The mechanical deformation processes are associated coating failure, similar to what is described in the External Corrosion mechanism, under the simultaneous action of External Stress that causes the plastic deformation of the pipeline (see the use of the AND gate—upper right part, **Figure 3**). The corrosion processes are activated when both the failure of the cathodic protection system and

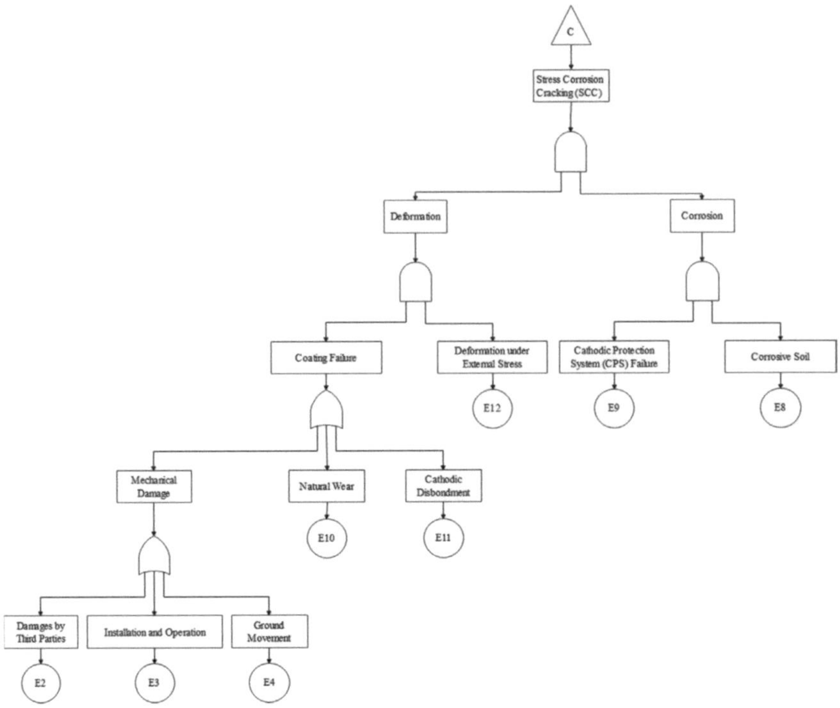

Figure 4.
Failure mechanism—stress corrosion cracking (SCC).

the presence of a corrosive environment occur simultaneously (see the use of the AND gate—upper left part, **Figure 4**). The transfer symbol C (in the upper left part of **Figure 3**) indicates where this branch inserts into the FTA presented in **Figure 1** (see the central right region of **Figure 1**).

The SCC mechanism can be divided into three main stages: initiation, propagation, and fracture [25, 47, 55]. In the initiation phase, surface defects or small cracks serve as stress concentration points, where the adsorption of corrosive ions, such as chlorides (for example), weakens the metallic bonds. In the propagation phase, the crack expands due to the simultaneous action of mechanical stress and electrochemical reactions that weaken the material along the crack front. Localized anodic dissolution occurs at the crack tip, where stress is the highest, facilitating metal oxidation. Additionally, cyclic loading due to gas pressure variations can accelerate crack propagation. When the crack reaches a critical size, the remaining stress may be sufficient to cause a catastrophic fracture of the pipeline [47, 55, 56].

The SCC mechanism in gas pipelines is a complex process influenced by a combination of mechanical, environmental, and material factors [25, 47, 55]. SCC is a degradation phenomenon that occurs in metallic materials, such as those used in pipelines, when a combination of mechanical stress and a specific corrosive environment causes the material to fail by fracture. This process can be particularly detrimental to pipelines due to the risk of gas leaks or catastrophic failures. A detailed understanding of the SCC process in pipelines, including the factors that influence the phenomenon, the mechanisms involved, and the mitigation practices, is essential for preventing such failures [48, 49, 56].

Mechanical stress is one of the main influencing factors of SCC in pipelines [50, 51]. This stress can be applied as a result of internal gas pressure, external loads, or residual stresses generated during the manufacturing or welding processes. Residual stresses, introduced during fabrication and installation, such as in welding, bending, and heat treatments, also play a significant role. Furthermore, the corrosive environment is crucial for SCC occurrence. The presence of moisture and contaminants, such as chlorides, sulfates, and carbonates, frequently found in the soil surrounding the pipelines, creates an electrolytic medium conducive to corrosion. The concentration of these aggressive ions can vary significantly depending on environmental conditions and geographic location [47, 50, 51].

The pipeline material also influences susceptibility to SCC [48]. The chemical composition, especially the presence of certain alloying elements, can make the steel more vulnerable. For example, high carbon content and impurities such as phosphorus and sulfur are particularly detrimental. Additionally, the material's microstructure, including the granular structure and the presence of secondary phases, can affect corrosion and cracking resistance [48, 53, 54].

To mitigate and prevent SCC, several strategies can be implemented. In pipeline design and fabrication, selecting materials with high SCC resistance, such as low-alloy steels specifically developed for aggressive environments, is fundamental. Heat treatments, such as annealing, can reduce internal residual stresses and increase corrosion resistance. Environmental control is also essential; applying protective coatings, such as epoxy or polyethylene, prevents the metal from contacting the corrosive environment. Additionally, adequate drainage around the pipeline minimizes the presence of moisture and corrosive ions [48, 53, 55].

Regular monitoring and inspection are crucial for detecting cracks and defects before they become critical. Non-destructive techniques, such as ultrasound and corrosion under insulation inspection, are widely used. Stress monitoring, through the use of sensors, allows the detection of anomalous variations that may indicate the presence of SCC. Cathodic protection, by whether impressed current or sacrificial anodes, can be applied to polarize the pipeline and reduce the corrosion rate. Treating the soil around the pipeline by adding neutralizing agents helps reduce the concentration of corrosive ions [48, 51, 54, 56].

Figure 5 presents the development of the fault tree for the corrosion fatigue mechanism, based on the literature [24, 25, 57–61].

The corrosion fatigue mechanism in gas pipelines depends on the simultaneous action of mechanical deformation and corrosion processes, similar to SCC, as indicated by the AND gate in **Figure 5**. The mechanical deformation processes are related to coating failure, akin to what is described in the mechanisms of External Corrosion and Stress Corrosion Cracking, with the simultaneous action of cyclic External Stress (see the use of the AND gate—upper right part, **Figure 5**). The corrosion processes are activated by the simultaneous failure of the cathodic protection system and the presence of a corrosive environment (see the use of the AND gate—upper left part, **Figure 5**). The transfer symbol D (in the upper left part of **Figure 5**) indicates where this branch inserts into the FTA presented in **Figure 1** (see the central right region of **Figure 1**).

Corrosion fatigue is a process that occurs when a metallic material, subjected to cyclic stresses, is also exposed to a corrosive environment [24, 25, 59]. The combination of these factors leads to accelerated material degradation, resulting in failures that can be catastrophic, especially in critical infrastructures like pipelines [59, 61].

The corrosion fatigue mechanism in pipelines begins with the application of cyclic stresses, which can originate from variations in the internal pressure of the

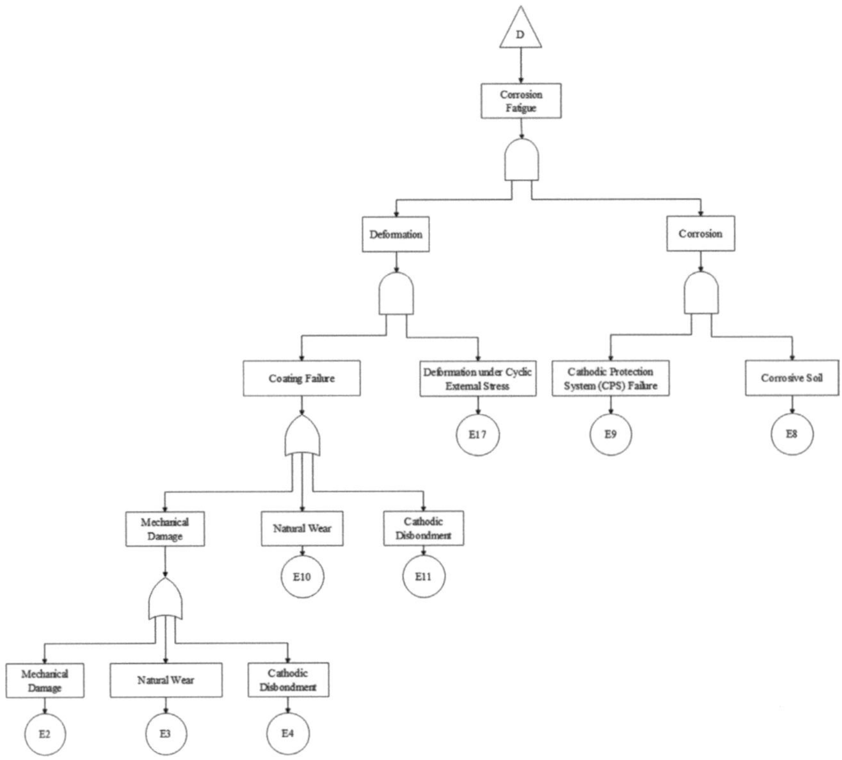

Figure 5.
Failure mechanism—corrosion fatigue.

transported gas or from external forces such as soil movement and operational loads. These cyclic stresses cause microcracks on the material's surface, serving as stress concentration points. In a corrosive environment, these points become even more vulnerable because the presence of corrosive agents, such as chlorides, sulfates, and other aggressive ions, accelerates the nucleation and growth of these cracks [57–59].

The interaction between stress cycles and the corrosive environment is synergistic. Each load and unload cycle introduce new microcracks or propagates existing ones, while corrosion progressively weakens the edges of these cracks. This weakening occurs through anodic dissolution mechanisms, where the metal at the crack edges is oxidized, reducing the material's strength and facilitating crack propagation. This continuous process leads to a reduction in the material's lifespan and can result in unexpected failures [24, 57–59].

The corrosive environment plays a critical role in the corrosion fatigue process. The presence of moisture and contaminants, frequently found in the soil surrounding pipelines, creates an electrolytic medium that promotes corrosion. Temperature also significantly influences the corrosion rate and its interaction with fatigue. In regions with large thermal variations, the phenomenon of thermal expansion and contraction can exacerbate the effects of cyclic stresses, further aggravating the corrosion fatigue process [25, 58, 59, 61].

Furthermore, the microstructure of the pipeline material is a determining factor in susceptibility to corrosion fatigue [60]. The chemical composition, heat treatments, and the history of residual stresses influence the material's resistance to crack nucleation and propagation. Materials with fine and homogeneous grains tend to have higher resistance to fatigue and corrosion. Conversely, impurities and inclusions can act as crack initiation points, reducing the pipeline's durability [57, 60].

To mitigate corrosion fatigue in pipelines, various strategies can be adopted. Firstly, selecting materials with high resistance to corrosion and fatigue is fundamental. High-strength steels, special metallic alloys, and surface treatments can significantly increase the durability of pipelines. The use of protective coatings, such as epoxy or polyethylene, prevents direct contact of the metal with the corrosive environment, reducing the corrosion rate [24, 25, 59, 61].

Continuous monitoring and regular inspection of pipelines are essential for the early detection of cracks and failures. Non-destructive techniques such as ultrasound, radiography, and visual inspection are widely used to assess structural integrity and identify potentially problematic areas. Implementing sensors to monitor stresses and detect anomalous variations is also an effective practice to predict and prevent failures [60].

Additionally, cathodic protection can be applied to reduce the corrosion rate by electrically polarizing the pipeline. Cathodic protection methods include the use of sacrificial anodes or the application of an impressed current. Finally, environmental management, such as controlling the soil chemistry around pipelines, can minimize the concentration of corrosive ions and reduce the risks of corrosion fatigue [57, 59, 61].

Figure 6 presents the development of the fault tree for the hydrogen embrittlement mechanism, based on the literature [11, 62–70].

The mechanism of hydrogen embrittlement (HE) in gas pipelines can develop from either the internal or external surface of the pipeline (note the OR gate at the top of **Figure 6**). The internal variant of hydrogen embrittlement (top right of **Figure 6**) is conditioned by the simultaneous action (see the use of the AND gate) of the ineffectiveness of the corrosion inhibitor (E5) in a corrosive fluid environment, specially related to H_2S concentration (E8), with a pipeline material possessing metallurgical characteristics that favor hydrogen permeation (E13). The external variant (right of **Figure 6**) involves the simultaneous occurrence of coating failure (as previously described in the external corrosion mechanism) in a pipeline with metallurgical characteristics conducive to hydrogen permeation (E13) combined with a scenario of cathodic overprotection, which can be exacerbated by certain environmental characteristics, especially soil pH (E14) or the presence of sulfate-reducing bacteria (SRB) (E15). The transfer symbol E (top left of **Figure 6**) indicates where this branch inserts into the Fault Tree Analysis (FTA) presented in **Figure 1** (see right, central region of **Figure 1**).

The mechanism of hydrogen embrittlement (HE) in pipelines is a phenomenon that occurs due to the interaction between hydrogen and the metallic material of the pipeline, resulting in a significant degradation of the material's mechanical properties. This process can lead to catastrophic failures, compromising the integrity and safety of pipelines [63, 63]. The interaction of hydrogen with the metallic material can occur through various mechanisms, including the formation of hydrides, the recombination of atomic hydrogen to form molecular hydrogen in internal traps, and direct interference with the metal's crystalline structure. These processes result in several adverse phenomena, such as reduced tensile strength, decreased ductility, and increased susceptibility to rapid crack propagation [11, 64].

Inside pipelines, hydrogen can be introduced during normal operation, especially in pipelines transporting natural gas containing hydrogen impurities. The hydrogen permeation process in steel pipes involves several critical stages. Initially, hydrogen

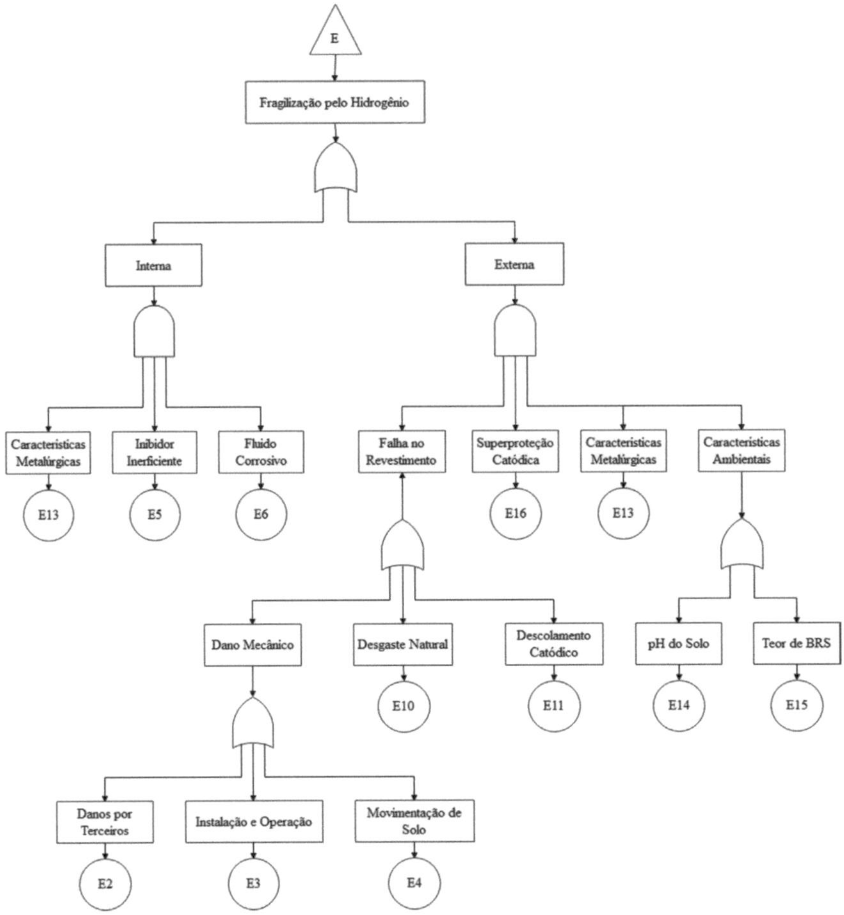

Figure 6.
Failure mechanism—hydrogen embrittlement.

dissociates into atoms on the pipeline's internal surface, a phenomenon occurring due to the high pressures and temperatures present during operation. These highly mobile hydrogen atoms begin to diffuse into the metallic material [65–67]. As hydrogen diffuses through the steel, it tends to migrate to areas of high stress, such as internal defects, inclusions, and regions with high residual stresses. These high-stress concentration areas act as traps for hydrogen atoms. Once trapped in these sites, hydrogen can recombine to form H_2 molecules, creating significant internal pressures that promote bubble formation. Over time, these bubbles can grow and coalesce, generating microfissures within the material [68].

The formation and propagation of these fissures reduce the ductility and toughness of the steel, making it more susceptible to fractures. This process is exacerbated by factors such as the material's microstructure, where the presence of fine grains or the distribution of secondary phases can influence hydrogen's ability to diffuse and accumulate. Additionally, residual stresses introduced during manufacturing and installation, such as those resulting from welding or inadequate heat treatments, can intensify hydrogen embrittlement by creating more high-stress locations where hydrogen can accumulate [63, 64, 68, 69]. The chemical composition of the steel also plays a crucial role. Elements such as sulfur and phosphorus can act as nucleation points for crack formation, while certain heat treatments can modify the steel's microstructure, increasing or decreasing its susceptibility to hydrogen embrittlement. In summary, internal hydrogen embrittlement is a multifaceted process involving hydrogen introduction, diffusion, accumulation, and eventual coalescence, leading to significant degradation of the material's mechanical properties and increasing the risk of catastrophic pipeline failures [66, 69].

Externally, the hydrogen embrittlement process can occur due to exposure to corrosive environments, such as wet soils contaminated with corrosive agents, and poorly controlled cathodic protection processes. While cathodic protection is essential to prevent widespread corrosion, it can result in hydrogen generation on the pipeline surface if not properly managed. Cathodic overprotection in pipelines, designed to prevent corrosion, can inadvertently contribute to the HE mechanism. Cathodic protection functions by applying an electric current to the pipeline, reducing the metal's tendency to corrode by creating a less favorable electrochemical environment for oxidation. However, when the applied current is excessive, overprotection occurs, resulting in the generation of atomic hydrogen on the steel pipeline surface. This highly reactive atomic hydrogen can easily diffuse into the metallic material [62, 64, 69].

The continuous process of hydrogen generation and diffusion, exacerbated by cathodic overprotection (application of a potential more negative than -1200 mV with respect to the Cu/CuSO4 reference electrode [71]), results in significant material embrittlement, reducing its ductility and toughness. Thus, while cathodic protection is essential for preventing corrosion, inadequate control can paradoxically increase the pipeline's susceptibility to hydrogen embrittlement, highlighting the importance of carefully monitoring and adjusting the applied current levels to avoid overprotection [63, 66, 68].

To mitigate hydrogen embrittlement in pipelines, several strategies are adopted. Firstly, selecting materials resistant to hydrogen is crucial. Low-alloy steels with controlled impurity levels, such as sulfur and phosphorus, are preferable, as these impurities can act as nucleation points for hydrogen-induced crack formation. Proper heat treatments, such as quenching and tempering, can improve resistance to hydrogen embrittlement by modifying the steel's microstructure to a more resistant configuration [68, 69]. Cathodic protection must be carefully controlled to avoid excessive hydrogen generation. This includes regular monitoring of protection potentials (application of a potential more negative than -1200 mV with respect to the Cu/$CuSO_4$ reference electrode [71]) and proper maintenance of the cathodic protection system. Additionally, high-quality protective coatings, such as epoxy or polyethylene, are essential to prevent direct metal contact with the corrosive environment and minimize hydrogen absorption [65–67].

Continuous monitoring and regular inspection of pipelines are crucial to detect signs of hydrogen embrittlement before catastrophic failure occurs. Advanced inspection techniques, such as acoustic emission tomography and electrochemical impedance spectroscopy, can be used to identify hydrogen presence and assess material integrity [66, 69, 72].

3. Proposal of a model to determine the global failure probability of the gas pipeline system using quantitative fault tree analysis (QFTA)

The proposed model integrates data from various sources, including historical failure records, real-time monitoring systems and/or periodic inspections, and environmental impact assessments through the technique of Quantitative Fault Tree Analysis (QFTA). By employing simple (yet robust) quantitative techniques from the field of Reliability Engineering, it is possible to dynamically update the fault tree parameters, ensuring that the model remains responsive to changing operational conditions. Furthermore, the approach includes a sensitivity analysis, performed through the calculation of Importance Measures, to identify the most critical factors influencing system reliability, allowing for targeted interventions that can significantly reduce the likelihood of catastrophic failures.

The initial stage of QFTA application involves generating a list of minimal cut sets, which offers essential qualitative insights. The model's accuracy was tested using Monte Carlo simulation, comparing the results obtained with other results from the literature and with results from various international databases for pipeline accident/incident control. Subsequently, importance measure analyses were conducted, all based on the minimal cut sets, to evaluate the criticality of the basic events involved in the model.

Minimal cut sets (MCS) are a group of sets that consist of the smallest combinations of basic events that result in the occurrence of the top event [15]. They represent all the ways in which basic events cause the top event and can be obtained by simplifying the Boolean algebra expressions derived from the quantitative fault tree analysis illustrated in **Figure 1**, which are listed in **Table 1**.

Basic event	Description
E1	Stray Current Corrosion
E2	Third-Party Damage
E3	Installation and Operation
E4	Ground Movement
E5	Ineffective Inhibitor
E6	Corrosive Fluid
E7	Water Accumulation Location
E8	Corrosive Soil
E9	Cathodic Protection Failure (application of a potential less negative than −850 mV vs. Cu/CuSO4 reference electrode [71])
E10	Natural Wear (Coating)
E11	Cathodic Disbondment (Coating)
E12	External Stress Deformation
E13	Metallurgical Characteristics (pipeline material)
E14	Soil pH
E15	Sulfate-Reducing Bacteria—SRB Content
E16	Cathodic Overprotection (application of a potential more negative than −1200 mV vs. Cu/CuSO4 reference electrode [71])
E17	Deformation under Cyclic External Stress

Table 1.
*Basic events of the FTA (**Figure 1**).*

Observing the correlations between the basic events in the FTA (**Figure 1**), it is possible to identify 31 cut sets (CS) listed in **Table 2**, along with their Boolean algebraic representation and described in relation to the basic events they comprise.

The minimal cut sets can be obtained by applying Boolean Algebra theorems to simplify these expressions [72], resulting in the 11 minimal cut sets (MCS) listed in **Table 3**, indicating their order, along with their Boolean algebraic representation and described in relation to the basic events they comprise.

The results of this minimal cut sets reduction indicate that mechanical damage (caused by third parties during installation and operation phases or due to ground movement), internal corrosion (with or without simultaneous application of external, constant, or cyclic stress), stray current corrosion, external corrosion (also with or without simultaneous application of external, constant, or cyclic stress), and cracking due to hydrogen embrittlement are the shortest and most direct pathways leading to gas release incidents in pipelines, through perforation or rupture.

This observation is consistent with data observed by researchers [73–77] in accident/incident databases (especially from the European Gas Pipeline Incident Data Group—EGIG, PetroChina Natural Gas & Pipeline Company—PNGPC, Canada Energy Regulator—CEA, and USA DOT Pipeline and Hazardous Materials Safety Administration—PHMSA), which indicate that gas pipeline infrastructure failures during the period 2014–2023 were most commonly caused by: corrosion (18%), material/weld/equipment failure (35%), excavation damage (18%), incorrect operation (8%), other damages from external forces (5%), and damages from natural forces (3%), resulting in the leakage of products contained within the gas pipelines.

The minimal cut sets represented in **Table 4** can be expressed by a Boolean Algebra equation (Eq. (1)) that allows estimating the overall probability of gas release failure in the gas pipeline (failure rate—λ) from the probability values of each basic event occurring independently:

$$\begin{aligned} PoF = {} & E1 + E2 + E3 + E4 + E5E6 + E8E9E10 + E8E9E11 + E10E13E14E16 \\ & + E11E13E14E16 + E10E13E15E16 + E11E13E15E16 \end{aligned} \tag{1}$$

The model above can be validated using the methodology proposed in **Figure 7**.

As shown in **Figure 7**, the methodology for validating the mathematical model for calculating the failure probability of a gas pipeline is based on the use of Monte Carlo simulation, which is a robust and widely used technique for modeling uncertainty in complex systems [92–97]. Initially, the pipeline failure probability will be determined through Monte Carlo simulations with 10.000 iterations (equivalent, e.g., to approximately 27.4 years of operation, or 78% of the average project life of gas pipelines). This approach allows the generation of a large amount of data, providing a more accurate and reliable estimate of the failure probability [92, 94, 96]. After performing the 10.000 iterations, the results will be expressed in terms of mean and standard deviation, providing a clear view of the distribution of simulated values [93, 95, 97].

To ensure the robustness of the results, the Monte Carlo simulations will be repeated 100 times [92, 94, 96]. Each of these 100 simulations will be analyzed individually, calculating the mean and standard deviation of the failure probability for each set of 10.000 iterations. This repetition captures the intrinsic variability of the simulation process and ensures that the results are not influenced by random fluctuations specific to a single simulation [93, 95, 97].

At the end of the 100 simulation rounds, the results will be consolidated to obtain a final estimate of the pipeline's failure probability. This estimate will be given by the

Cut set (CS)	Boolean algebraic representation	Description
CS1	E1	Failure due to Stray Current Corrosion
CS2	E2	Failure due to Third-Party Damage
CS3	E3	Failure due to Mechanical Damage during Installation and/or Operation
CS4	E4	Failure due to Mechanical Damage from Ground Movement
CS5	E5E6	Failure due to Internal Corrosion originating from ineffective inhibitor in the presence of a corrosive fluid
CS6	E2E8E9	Failure due to External Corrosion caused by damage to the coating by third-party action associated with ineffective cathodic protection[a] in corrosive soil
CS7	E3E8E9	Failure due to External Corrosion caused by mechanical damage to the coating during installation and/or operation phases associated with ineffective cathodic protection[a] in corrosive soil
CS8	E4E8E9	Failure due to External Corrosion caused by mechanical damage to the coating from ground movement associated with ineffective cathodic protection[a] in corrosive soil
CS9	E8E9E10	Failure due to External Corrosion caused by natural wear (aging) of the coating associated with ineffective cathodic protection[a] in corrosive soil
CS10	E8E9E11	Failure due to External Corrosion caused by cathodic disbondment of the coating associated with ineffective cathodic protection[a] in corrosive soil
CS11	E5E6E7	Failure due to Internal Corrosion caused by ineffective inhibitor in the presence of a corrosive fluid at a water accumulation location in the pipeline
CS12	E2E8E9E12	Failure due to Stress Corrosion Cracking (SCC) caused by damage to the coating by third-party action associated with ineffective cathodic protection[a] in corrosive soil under simultaneous external stress
CS13	E3E8E9E12	Failure due to Stress Corrosion Cracking (SCC) caused by mechanical damage to the coating during installation and/or operation phases associated with ineffective cathodic protection[a] in corrosive soil under simultaneous external stress
CS14	E4E8E9E12	Failure due to Stress Corrosion Cracking (SCC) caused by mechanical damage to the coating from ground movement associated with ineffective cathodic protection[a] in corrosive soil under simultaneous external stress
CS15	E8E9E10E12	Failure due to Stress Corrosion Cracking (SCC) caused by natural wear (aging) of the coating associated with ineffective cathodic protection[a] in corrosive soil under simultaneous external stress
CS16	E8E9E11E12	Failure due to Stress Corrosion Cracking (SCC) caused by cathodic disbondment of the coating associated with ineffective cathodic protection[a] in corrosive soil under simultaneous external stress
CS17	E2E8E9E17	Failure due to Fatigue Corrosion caused by damage to the coating by third-party action associated with ineffective cathodic protection[a] in corrosive soil under simultaneous cyclic external stress
CS18	E3E8E9E17	Failure due to Fatigue Corrosion caused by mechanical damage to the coating during installation and/or operation phases associated with ineffective cathodic protection[a] in corrosive soil under simultaneous cyclic external stress

Cut set (CS)	Boolean algebraic representation	Description
CS19	E4E8E9E17	Failure due to Fatigue Corrosion caused by mechanical damage to the coating from ground movement associated with ineffective cathodic protection[a] in corrosive soil under simultaneous cyclic external stress
CS20	E8E9E10E17	Failure due to Fatigue Corrosion caused by natural wear (aging) of the coating associated with ineffective cathodic protection[a] in corrosive soil under simultaneous cyclic external stress
CS21	E8E9E11E17	Failure due to Fatigue Corrosion caused by cathodic disbondment of the coating associated with ineffective cathodic protection[a] in corrosive soil under simultaneous cyclic external stress
CS22	E2E13E14E16	Failure due to Hydrogen Embrittlement (HE) cracking caused by mechanical damage to the coating from third-party action in a pipeline with metallurgical characteristics favoring hydrogen permeation under a situation of hydrogen generation due to cathodic overprotection[b] potentiated by Soil pH
CS23	E3E13E14E16	Failure due to Hydrogen Embrittlement (HE) cracking caused by mechanical damage to the coating during installation and/or operation phases in a pipeline with metallurgical characteristics favoring hydrogen permeation under a situation of hydrogen generation due to cathodic overprotection[b] potentiated by Soil pH
CS24	E4E13E14E16	Failure due to Hydrogen Embrittlement (HE) cracking caused by mechanical damage to the coating from ground movement in a pipeline with metallurgical characteristics favoring hydrogen permeation under a situation of hydrogen generation due to cathodic overprotection[b] potentiated by Soil pH
CS25	E10E13E14E16	Failure due to Hydrogen Embrittlement (HE) cracking caused by mechanical damage to the coating from soil movement in a pipeline with metallurgical characteristics favoring hydrogen permeation under a situation of hydrogen generation due to cathodic overprotection[b] potentiated by Soil pH
CS26	E11E13E14E16	Failure due to Hydrogen Embrittlement (HE) cracking caused by cathodic disbondment of the coating in a pipeline with metallurgical characteristics favoring hydrogen permeation under a situation of hydrogen generation due to cathodic overprotection[b] potentiated by Soil pH
CS27	E2E13E15E16	Failure due to Hydrogen Embrittlement (HE) cracking caused by mechanical damage to the coating from third-party action in a pipeline with metallurgical characteristics favoring hydrogen permeation under a situation of hydrogen generation due to cathodic overprotection[b] potentiated by sulfate-reducing bacteria (SRB) content in the soil
CS28	E3E13E15E16	Failure due to Hydrogen Embrittlement (HE) cracking caused by mechanical damage to the coating during installation and/or operation phases in a pipeline with metallurgical characteristics favoring hydrogen permeation under a situation of hydrogen generation due to cathodic overprotection[b] potentiated by sulfate-reducing bacteria (SRB) content in the soil
CS29	E4E13E15E16	Failure due to Hydrogen Embrittlement (HE) cracking caused by mechanical damage to the coating from ground movement in a pipeline with metallurgical characteristics favoring hydrogen permeation under a situation of hydrogen generation due to cathodic overprotection[b] potentiated by sulfate-reducing bacteria (SRB) content in the soil

Cut set (CS)	Boolean algebraic representation	Description
CS30	E10E13E15E16	Failure due to Hydrogen Embrittlement (HE) cracking caused by natural wear (aging) of the coating in a pipeline with metallurgical characteristics favoring hydrogen permeation under a situation of hydrogen generation due to cathodic overprotection[b] potentiated by sulfate-reducing bacteria (SRB) content in the soil
CS31	E11E13E15E16	Failure due to Hydrogen Embrittlement (HE) cracking caused by cathodic disbondment of the coating in a pipeline with metallurgical characteristics favoring hydrogen permeation under a situation of hydrogen generation due to cathodic overprotection[b] potentiated by sulfate-reducing bacteria (SRB) content in the soil

[a]The Cathodic Protection System (CPS) will be ineffective when applying a potential less negative than −850 mV versus a Cu/CuSO4 reference electrode [71].[b]Cathodic overprotection occurs when applying a potential more negative than −1200 mV versus a Cu/CuSO4 reference electrode [71].

Table 2.
*Cut sets (CS) of the FTA (**Figure 1**).*

average of the means of the 100 simulations, and the standard deviation will be used to quantify the uncertainty associated with this measure [93, 95, 97]. The mean value obtained will then be compared with reference values available in the literature and relevant databases. This comparison will allow the assessment of the accuracy and validity of the developed mathematical model, ensuring that it aligns with the established standards in the field of Reliability Engineering.

To model the randomness of basic events, the exponential distribution was chosen [98–100]. The exponential distribution assumes that the time between failures is independent, meaning that the occurrence of one failure does not influence the probability of future failures [98]. In many cases, pipeline failures, such as ruptures due to accidental damage or leaks caused by corrosion, can be modeled as independent events over time [99, 100]. This temporal independence of failures justifies the use of the exponential distribution, which is characterized by a constant failure rate (λ) [99, 100].

The constant failure rate property of the exponential distribution is suitable for situations where the pipeline operates under relatively stable conditions, and the failure probability is approximately constant during the analysis period. For example, in a well-maintained pipeline operated within normal parameters, the failure rate can remain relatively stable, simplifying modeling and statistical analysis [98]. Moreover, the mathematical simplicity of the exponential distribution facilitates computational implementation and statistical analysis, allowing efficient integration with Monte Carlo simulation and other reliability analysis tools [95].

To obtain reference values for model validation and selection of the most appropriate failure rate (λ) for simulations, a literature review and official pipeline accident/incident databases were consulted, resulting in the values presented in **Table 4**.

Based on these results, it was decided to adopt a failure rate (λ) value of $0.577 \times 10^{-3}\ km^{-1}.\ year^{-1}$ to model the randomness of basic events. This value is in the upper range of the sample (the third highest value in the range of 0.03 to $6.46 \times 10^{-3}\ km^{-1}.\ year^{-1}$) and is chronologically the most recent.

Following the flowchart shown in **Figure 7**, Monte Carlo simulations were performed, resulting in the graphs shown in **Figures 8** and **9** and the consolidated

Minimal cut set (MCS)	Order	Boolean algebraic representation	Description
MCS1	1[a]	E1	Failure due to Stray Current Corrosion
MCS2	1[a]	E2	Failure due to Third-Party Damage
MCS3	1[a]	E3	Failure due to Mechanical Damage during Installation and/or Operation Phases
MCS4	1[a]	E4	Failure due to Mechanical Damage from Ground Movement
MCS5	2[a]	E5E6	Failure due to Internal Corrosion caused by ineffective inhibitors in the presence of a corrosive fluid
MCS6	3[a]	E8E9E10	Failure due to External Corrosion caused by mechanical damage to the coating from soil movement associated with ineffective cathodic protection[a] in corrosive soil
MCS7	3[a]	E8E9E11	Failure due to External Corrosion caused by damage to the coating from natural wear (aging) associated with ineffective cathodic protection[a] in corrosive soil
MCS8	4[a]	E10E13E14E16	Failure due to Hydrogen Embrittlement (HE) cracking caused by mechanical damage to the coating from natural wear (aging) in a pipeline with metallurgical characteristics favoring hydrogen permeation under a situation of hydrogen generation due to cathodic overprotection[b] potentiated by Soil pH
MCS9	4[a]	E11E13E14E16	Failure due to Hydrogen Embrittlement (HE) cracking caused by cathodic disbondment of the coating in a pipeline with metallurgical characteristics favoring hydrogen permeation under a situation of hydrogen generation due to cathodic overprotection[b] potentiated by Soil pH
MCS10	4[a]	E10E13E15E16	Failure due to Hydrogen Embrittlement (HE) cracking caused by mechanical damage to the coating from natural wear (aging) in a pipeline with metallurgical characteristics favoring hydrogen permeation under a situation of hydrogen generation due to cathodic overprotection[b] potentiated by sulfate-reducing bacteria (SRB) content in the soil
MCS11	4[a]	E11E13E15E16	Failure due to Hydrogen Embrittlement (HE) cracking caused by cathodic disbondment of the coating in a pipeline with metallurgical characteristics favoring hydrogen permeation under a situation of hydrogen generation due to cathodic overprotection[b] potentiated by sulfate-reducing bacteria (SRB) content in the soil

[a]The Cathodic Protection System (CPS) will be ineffective when applying a potential less negative than −850 mV versus a Cu/CuSO4 reference electrode [71].[b]Cathodic overprotection occurs when applying a potential more negative than −1200 mV versus a Cu/CuSO4 reference electrode [71].

Table 3.
*Minimal cut sets (MCS) of the FTA (**Figure 1**).*

values are expressed in **Table 5**. The Monte Carlo simulations were implemented using MATLAB software.

Considering the mean and standard deviation values obtained in the simulation process according to the flowchart in **Figure 7** for the gas pipeline gas release failure probability, and calculating the uncertainty associated with the adopted exponential distribution according to the Guide to the Expression of Uncertainty in Measurement

Reference	Year or period	Failure rate [10^{-3} km^{-1}. year^{-1}]	Type of reference
Hong et al. [23]	2023	5.77	Article proposing a model for estimating failure rate (λ)
CONCAWE[a] (Europa) [78]	2020	0.28	Database
DoT[b] (USA) [79]	2020	0.16	Database
EGIG[c] (Europa) [80]	2020	0.089	Database
Golovach et al. [81]	2020	0.2537-0.5547	Article referencing Databases
Li et al. [82]	2019	0.046	Article referencing Databases
Li et al. [83]	2018	0.065	Article referencing Databases
Dai et al. [84]	2017	0.4-0.6	Article referencing Databases
Lever et al. [85]	2017	0.03-0.05	Article referencing Databases
Cunha [86]	2012	0.23	Article referencing Databases
Shahriar et al. [87]	2012	6.25	Article proposing a model for estimating failure rate (λ)
Dawotola et al. [83]	2011	0.79-2.18	Article proposing a model for estimating failure rate (λ)
Caleyo et al. [88]	2008	3	Article proposing a model for estimating failure rate (λ)
Yuhua and Datao [89]	2005	6.46	Article proposing a model for estimating failure rate (λ)
DoT[b] (EUA) [79]	2001–2006	0.34	Database
EGIGc (Europa) [80]	1995–2001	0.21	Database
CER[d] (Canadá) [90]	2000–2008	0.10	Database
EGIG[c] (Europa) [80]	2010–2019	0.095	Database
PHMSA[e] (EUA) [91]	2010–2019	0.11	Database
PHMSA[e] (EUA) [91]	2001–2020	0.1 – 0.5	Database
EGIG[c] (Europa) [80]	1970–2001	0.44	Database
TRANSPETRO[f] (Brasil) [86]	1978–2010	0.36	Database
CONCAWE[a] (Europa) [78]	1971–2006	0.38	Database
UKOPA[g] (Reino Unido) [81]	1962–2019	0.23	Database

[a]CONCAWE—Conservation of Clean Air and Water in Europe.[b]DoT—Department of Transportation (USA).[c]EGIG—European Gas Pipeline Incident Data Group.[d]CER—Canada Energy Regulator.[e]PHMSA—Pipeline and Hazardous Materials Safety Administration (USA).[f]TRANSPETRO—Petrobras Transporte S.A.[g]UKOPA—United Kingdom Onshore Pipeline Operators' Association.

Table 4.
*Reference values for failure rate (λ) (**Figure 7**).*

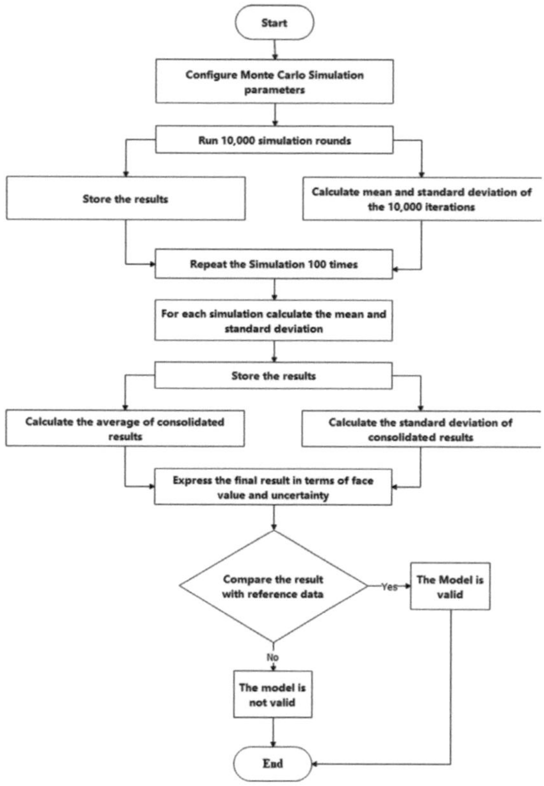

Figure 7.
Validation of the proposed model (flowchart).

(GUM) [101], it is possible to attribute to the randomly obtained failure rate (λ) by the proposed model the value of $(1.2373 \pm 0.0058) \times 10^{-3}$ km^{-1}. year^{-1}.

Graphically analyzing the results (**Figures 8** and **9**), it is evident that both the mean and standard deviation of the simulations exhibit homoscedasticity and uniformity, indicating high model convergence within the 95% confidence interval.

Figure 10 presents a graphical representation of the result achieved by applying the proposed model in relation to the reference values from **Table 4**.

In **Figure 10**, the reference values from literature data are represented with markers in black, highlighting the authors, while the reference values from official accident/incident databases are represented with markers in blue, without specific labels. The value obtained from the application of the proposed model is highlighted with a marker in red.

It can be observed that the failure rate obtained with the proposed model closely aligns with values recorded in the databases across various periods of the reference research [78–81, 90, 91], as well as with values cited by researchers in articles referencing databases [81–86], placing historical failure rates in the range of 0.046 to 0.6×10^{-3} km^{-1}. year^{-1}. Similarly, the simulation results using the proposed model yielded a failure rate lower than those achieved by other studies proposing models with the similar objectives [23, 83, 87–89, 102, 103], which ranged from 2.18 to 6.46×10^{-3} km^{-1}. year^{-1}.

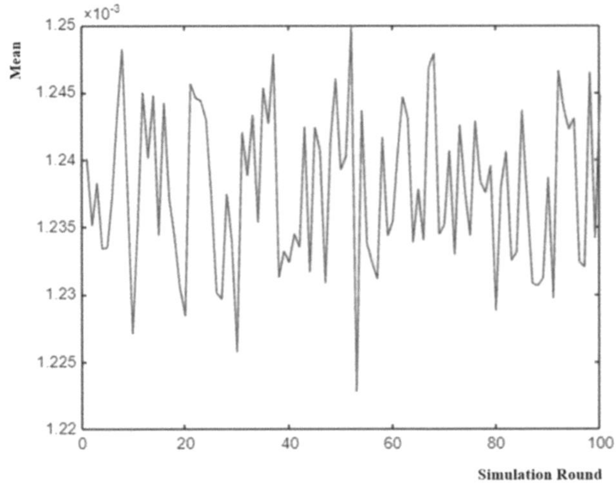

Figure 8.
Results of Monte Carlo simulations for the mean parameter.

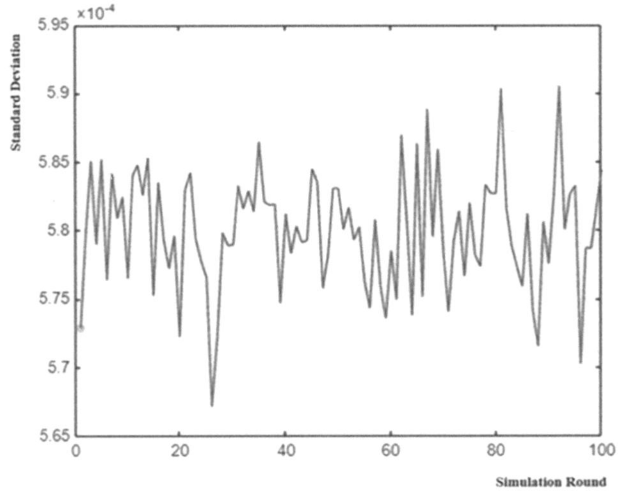

Figure 9.
Results of Monte Carlo simulations for the standard deviation parameter.

Parameter	Value
Mean	0.00123730
Standard deviation	0.00057969

Table 5.
Results of Monte Carlo simulations.

Therefore, it is possible to affirm that the model is valid for estimating failure rates, as the result is conservative compared to the historically recorded level of failure rates, without overestimating the results as observed in other models studied as

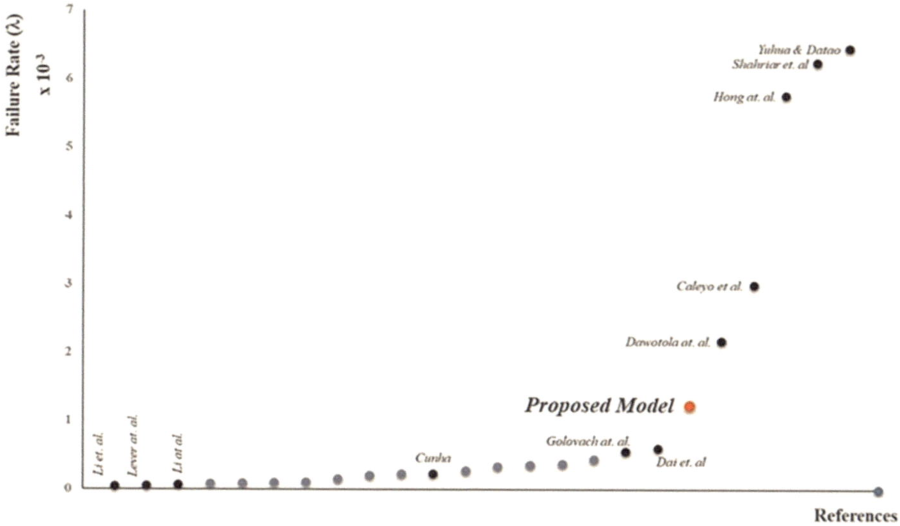

Figure 10.
*Failure rate (λ) obtained with the model compared to the reference failure rate (λ) (**Table 4**).*

references. Thus, the application of the model demonstrates the ability to provide accurate estimates for the application of resulting failure rates in Risk-Based Maintenance and Inspection practices, without the inconvenience of overestimating these values, which would imply extraordinary costs for control and mitigation actions, often exceeding what is actually required by the assessed risk.

Another positive aspect to highlight regarding the results of the proposed model is the methodology applied in modeling that allows for the calculation and expression of uncertainty based on the Guide to the Expression of Uncertainty in Measurement (GUM) [101]. This characteristic is not shared by other studies analyzed during the reference research phase. Uncertainty information is crucial for more in-depth studies in the field of Reliability Engineering when using model results in predictive maintenance techniques, maintainability studies, and advanced quantitative risk analysis.

4. Criticality analysis of basic failure events in pipelines through importance measures

In order to enhance the QFTA with an assessment of the criticality of basic events that lead to gas release failures in pipelines, importance measures can be utilized to determine their criticality. This approach has the potential to provide essential information for Reliability Engineering teams and other responsible parties to act more decisively in preventing failures, especially catastrophic ones, or at least work more efficiently toward their mitigation [2].

The structural integrity of gas pipelines involves the continuous assessment and maintenance of fluid transport systems, such as oil and gas, to prevent catastrophic failures that can result in economic losses and environmental impacts. Importance measures are analytical tools that help identify critical components within these systems, allowing for the prioritization of inspections, maintenance, and improvements [11, 104].

Importance measures provide a quantitative way to assess the role of each component in the overall system reliability [2, 105]. They indicate which components have the greatest influence on the system's failure probability and which are most critical for safe and efficient operation. These measures are fundamental in risk analysis and decision-making for the allocation of maintenance resources [106, 107].

In the context of pipeline structural integrity management, importance measures can be applied to prioritize inspections, identifying the most critical pipeline segments that require more frequent monitoring [2, 106]. Additionally, they are useful in maintenance planning, directing resources to components whose performance is most critical for the safe operation of the system. These measures also aid in scenario analysis, evaluating the impact of different maintenance and operation strategies, and in reliability improvement, identifying opportunities for incremental improvements in the reliability of the most important components [107].

Four commonly used importance measures are Birnbaum's Importance, Criticality Importance (often simply referred to as criticality), Event Sensitivity, and Fussell-Vesely Importance [107, 108].

Birnbaum's Importance (B_i) measures the potential contribution of a component or event i to the system failure [2, 106]. It can be calculated using Eq. (2):

$$B_i = \frac{\partial R}{\partial q_i} \tag{2}$$

where R is the system reliability and q_i is the success probability of component or event i.

Criticality Importance (CI_i) combines the failure probability of the component or event with the impact of that failure on the system [2, 106] and is calculated using Eq. (3):

$$CI_i = q_i.B_i \tag{3}$$

where q_i is the failure probability of the component or event, and B_i is Birnbaum's Importance.

Event Sensitivity (S_i) assesses the system's response to changes in the reliability of a specific component [2, 106]. This measure is given by Eq. (4):

$$B_i = \frac{\partial R}{\partial \lambda_i} \tag{4}$$

where λ_i is the failure rate of component or event i, and R is the system reliability.

Lastly, Fussell-Vesely Importance (FV_i) quantifies the contribution of a specific component to the system's total risk [2, 106]. It is defined using Eq. (5):

$$FV_i = \frac{Q_i}{Q_T} \tag{5}$$

where Q_i is the system failure probability due to component or event i, and Q_T is the total system failure probability.

In addition to these measures, there are two other important risk metrics: Risk Achievement Worth (RAW) and Risk Reduction Worth (RRW) [104, 105, 107, 108].

RAW_i is used to determine the increase in the total system risk if a given component or event is in the worst possible condition, that is, with a high failure rate or failing [104, 107]. It is calculated using Eq. (6):

$$RAW_i = \frac{Q_{falha\ i}}{Q_T} \tag{6}$$

where $Q_{failure\ i}$ is the system failure probability considering that component or event i is in a failed state, and Q_T is the current total system failure probability.

RRW_i measures the reduction in the total system risk if a specific component is in the best possible condition, that is, operating perfectly [104, 107]. It is calculated using Eq. (7):

$$RAW_i = \frac{Q_T}{Q_{sucesso\ i}} \tag{7}$$

where $Q_{success\ i}$ is the system failure probability considering that component or event i is operating perfectly.

Used together, these measures provide a comprehensive understanding of system vulnerabilities and guide decision-making for continuous reliability and safety improvements, especially in pipeline structural integrity management [105, 108].

To illustrate the application of these measures to pipeline structural integrity assessment, consider an example based on failure rate values obtained from the DoT—Department of Transportation (USA) database for the period 2001–2006 [79], which are detailed in **Table 6**.

Applying the values from **Table 6** to Eqs. (3)–(7) allows for the determination of the importance measures for each of the basic events leading to gas release failure in a pipeline, as listed in **Table 7** (by value) and **Table 8** (ranking).

Based on the results presented in **Tables 7** and **8**, it can be concluded that the failure events in Cathodic Protection System (CPS) Failure (E9), Cathodic Disbondment (coating) (E11), Soil Corrosivity (E8), Natural Wear (coating) (E10), and Cathodic Overprotection (E16) have the greatest impact on the system reliability concerning gas release failures. These events should be considered critical, emphasizing the need to focus on risk mitigation strategies and resource allocation to minimize their impact, thereby optimizing system reliability and safety more effectively.

Furthermore, it is noteworthy that the results of the event sensitivity measure Si also highlight Corrosion by Stray Current (E1), Water Accumulation Site (E7), Damage during Installation and Operation Phases (E3), Ground Movement (E4), External Stress Deformation (E12), and Metallurgical Characteristics of the pipeline (E13). Despite their low failure rates (λ), improvements or deteriorations in their reliability will have a more pronounced effect on the overall system reliability. Therefore, preventive control measures should be maintained or enhanced for these events.

5. Guidelines for the use of the proposed model in pipeline structural integrity management processes

Effective Pipeline Structural Integrity Management (PSIM) is an essential component for the safe and reliable operation of gas transportation grids. The continuous evolution of standards, technologies, and practices associated with PSIM contributes

Basic event	Description	Failure rate $[\text{x } 10^{-3}\text{ km}^{-1}.\text{ano}^{-1}]$
E1	Stray Current Corrosion	0.000011
E2	Third-Party Damage	0.000067
E3	Installation and Operation	0.000025
E4	Ground Movement	0.000029
E5	Ineffective Inhibitor	0.007100
E6	Corrosive Fluid	0.006200
E7	Water Accumulation Site	0.000015
E8	Corrosive Soil	0.012500
E9	Cathodic Protection System (CPS) Failure	0.016100
E10	Natural Wear (Coating)	0.012100
E11	Cathodic Disbondment (Coating)	0.013100
E12	External Stress Deformation	0.000047
E13	Metallurgical Characteristics	0.000047
E14	Soil pH	0.000125
E15	BRS Content	0.000121
E16	Cathodic Overprotection	0.007245
E17	Cyclical External Stress Deformation	0.000047

Table 6.
Failure rate (λ) by basic event [79].

to the ongoing improvement of the safety, efficiency, and sustainability of these critical systems [109].

Among the techniques and tools employed in PSIM are inspection and monitoring, risk assessment, modeling and simulation, preventive and corrective maintenance, and data management [8]. The application of quantitative and qualitative methods to assess the probability and consequences of failures helps prioritize maintenance actions [110]. Computational modeling can be used to predict pipeline behavior under different operational and environmental conditions, while condition-based maintenance strategies should include guidelines for repairs, replacements, and structural reinforcements. Integrated data management systems must be capable of collecting, storing, and analyzing relevant information about pipeline integrity [8, 109, 110].

The proposed model fits into this context as a technique to support actions within the scope of maintenance strategies focused on optimizing pipeline structural integrity based on maximum reliability, risk minimization, and life extension. This generates triggers for the correct application of actions within the scope of Risk-Based Inspection (RBI) and Reliability-Centered Maintenance (RCM).

To achieve this objective more efficiently, it is necessary to follow certain usage guidelines, as illustrated in the flowchart shown in **Figure 11**.

The foundation of applying the model lies in the continuous improvement of data collection systems, as data consistency is a crucial factor for accurately estimating the system's failure rate (λ) and determining the importance measures related to the basic events that need to be controlled. Incomplete, non-existent, or inaccurately recorded

Basic event	Description	B_i $[10^{-3}]$	FV_i	CI_i	S_i	RAW_i $[10^{-3}]$	RRW_i
E1	Stray Current Corrosion	0.0110	0.0607	0.0607	10.002	0.0110	0.0608
E2	Third-Party Damage	0.0670	0.3700	0.3700	10.001	0.0670	0.3700
E3	Installation and Operation	0.0250	0.1381	0.1381	10.002	0.0250	0.1381
E4	Ground Movement	0.0290	0.1602	0.1602	10.002	0.0290	0.1602
E5	Ineffective Inhibitor	71.521	39.2094	39.2094	0.9931	70.509	38.9380
E6	Corrosive Fluid	62,398	34.2392	34.2392	0.9940	61.627	34.0331
E7	Water Accumulation Site	0.0150	0.0828	0.0828	10.002	0.0150	0.0829
E8	Corrosive Soil	12.6605	69.0306	69.0306	0.9877	12.3460	68,1801
E9	Cathodic Protection System (CPS) Failure	16,3664	88.9114	88.9114	0.9841	15,8437	87.4958
E10	Natural Wear (Coating)	12.2504	66.8216	66.8216	0.9881	11,9558	66.0250
E11	Cathodic Disbondment (Coating)	13.2763	72.3441	72.3441	0.9871	12.9307	71.4093
E12	External Stress Deformation	0.0470	0.2596	0.2596	10.001	0.0470	0.2596
E13	Metallurgical Characteristics	0.0470	0.2596	0.2596	10.001	0.0470	0.2596
E14	Soil pH	0.1250	0.6903	0.6903	10.001	0.1250	0.6903
E15	BRS Content	0.1210	0.6682	0.6682	10.001	0.1210	0.6683
E16	Cathodic Overprotection	72.992	40.0101	40.0101	0.9929	71.938	39.7275
E17	Cyclical External Stress Deformation	0.0470	0.2596	0.2596	10.001	0.0470	0.2596

Note: Bi—Birnbaum Factor; FVi—Fussell-Vesely Factor; CIi—Critical Importance Factor; Si—Event Sensitivity; RAW$_i$—Risk Achievement Worth; RRW$_i$—Risk Reduction Worth.

Table 7.
Importance measures for the basic events in ***Table 6****.*

data are an unacceptable source of uncertainty in applying the model, potentially leading to incorrect estimations. From the perspective of RBI and RCM strategies being applied in the PSIM process, it is crucial to develop actions that promote the application of failure analysis and treatment techniques for all inspection and maintenance events in the gas pipeline. This ensures that each failure is correctly stratified not only by its final outcome but also by the basic events that contributed to the manifestation of the final effect.

Once the data is entered into the model and results for the failure rate (λ) and importance measures are obtained, the evaluation of these results must be technical and thorough, based on the most up-to-date standards, both external and internal to the installation manager. If the failure rate (λ) is unacceptable and/or the importance measures indicate the criticality of basic events, it is necessary to propose measures to eliminate, correct, or mitigate the non-conformities. At this stage, failure treatment must be swift, with timely responses and a thorough analysis of the causes and effects. The portfolio of RBI and RCM techniques and tools offers various means to execute failure treatment, which should be used judiciously to achieve maximum efficiency and to meet the objectives with optimal resource utilization.

Basic event	Description	B_i	FV_i	CI_i	S_i	RAW_i	RRW_i
E1	Stray Current Corrosion	17	17	17	1	17	17
E2	Third-Party Damage	10	10	10	8	10	10
E3	Installation and Operation	15	15	15	3	15	15
E4	Ground Movement	14	14	14	4	14	14
E5	Ineffective Inhibitor	6	6	6	12	6	6
E6	Corrosive Fluid	7	7	7	11	7	7
E7	Water Accumulation Site	16	16	16	2	16	16
E8	Corrosive Soil	3	3	3	15	3	3
E9	Cathodic Protection System (CPS) Failure	1	1	1	17	1	1
E10	Natural Wear (Coating)	4	4	4	14	4	4
E11	Cathodic Disbondment (Coating)	2	2	2	16	2	2
E12	External Stress Deformation	11	11	11	5	11	11
E13	Metallurgical Characteristics	11	11	11	5	11	11
E14	Soil pH	8	8	8	9	8	8
E15	BRS Content	9	9	9	9	9	9
E16	Cathodic Overprotection	5	5	5	13	5	5
E17	Cyclical External Stress Deformation	11	11	11	5	11	11

Note: B_i—Birnbaum Factor; FV_i—Fussell-Vesely Factor; CI_i—Critical Importance Factor; S_i—Event Sensitivity; RAW_i—Risk Achievement Worth; RRW_i—Risk Reduction Worth.

Table 8.
*Importance measures for the basic events in **Table 6** (Ranking).*

Once corrective actions and/or identified improvement opportunities are planned, they must be executed precisely, following the protocols defined for each action, respecting the timelines and scope outlined in the previous stage. After executing the corrective actions and implementing the identified improvements, a period of monitoring and control follows. During this period, there is usually a contraction in the standard inspection interval to identify any changes in the system's response to the executed actions, observing their effectiveness. Throughout this period, new data should be collected and fed back into the database, continuing the evaluation cycle. This establishes a continuous improvement process aimed at achieving maximum installation performance within strict reliability and risk management standards.

Finally, it is worth noting that the model can also be used during the design and pre-design phases of installations to strengthen the PSIM of facilities throughout their entire lifecycle. In these applications, the model can be utilized with simulation techniques, such as Monte Carlo simulation, based on data from expert panels, benchmarking with other similar installations, or official databases.

6. Final considerations

Considering the critical importance of gas pipelines in the transportation of energy resources, improving the reliability of these pipelines is essential to avoid costly

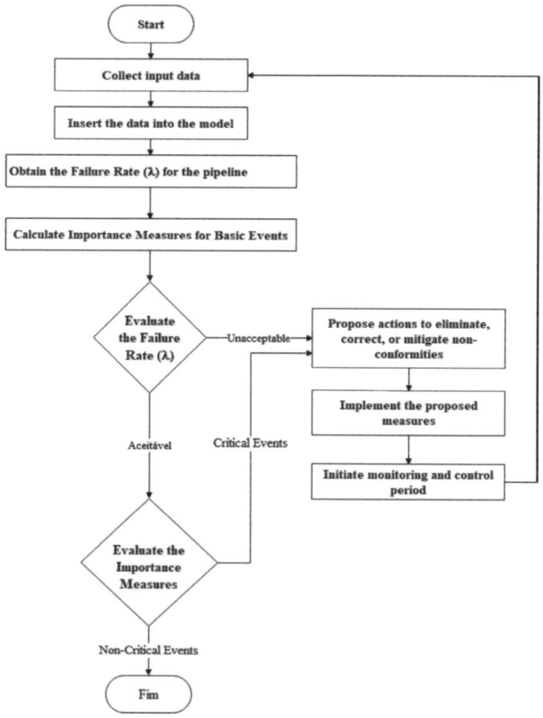

Figure 11.
Application of the proposed model (flowchart).

failures and environmental damage, and ensure safe operations. The innovative approach proposed in this article for managing pipeline integrity using the Minimum Cut Set (MCS) method and Importance Measures has developed a practical, straightforward, and effective model. This is evidenced when compared with the results from various official databases on incidents/accidents with such utilities, as well as with similar models proposed in the literature.

By constructing the fault tree for gas release, considering the failure modes of puncture and rupture and the different mechanisms involved, the Minimum Cut Set technique allowed for the identification of critical failure scenarios. At the same time, Importance Measures enabled the assignment of criticality to each event's contribution to the system's reliability. Prioritizing influential factors optimizes resources for inspection and maintenance, thereby improving the overall reliability of the pipelines.

Thus, by integrating Reliability Engineering principles, the study developed a comprehensive framework that can be used for the design, operation, and maintenance of pipeline networks. The proposed and validated model has the potential to significantly contribute to the knowledge of pipeline integrity management, offering practical solutions for the safety and sustainability of energy transportation infrastructure.

The straightforward and direct guidelines provided for applying the model ensure that it can be easily implemented by professionals working with PSIM and the entire scientific community dedicated to research in this field of knowledge.

Nomenclature list

AFA	applied failure analysis
AMPP	association for materials protection and performance
API	American Petroleum Institute
API	application programming interfaces
ASM	American Society for Metals
ASME	American Society of Mechanical Engineers
CIPS	close interval potential survey
CMS	corrosion management system
CPS	cathodic protection system
DCVG	direct current voltage gradient
DNV	Det Norske Veritas
DoT	department of transportation (USA)
FEA	finite element analysis
FMEA	failure modes and effects analysis
FTA	fault-tree analysis
GDP	gross domestic product
GIS	geographic information system
GUM	guide to the expression of uncertainty in measurement
HAZOP	hazard and operability analysis
HE	hydrogen embrittlement
ILI	in-line inspections
ISO	international organization for standardization
KPI	key performance indicators
MCDA	multi-criteria decision analysis
MSC	minimal set cuts
NACE	National Association of Corrosion Engineers (USA)
NBR	Brazilian Standard
NDT	non-destructive testing
NFT	near field testing
PGI	integrity management plan
PHMSA	Pipeline and Hazardous Materials Safety Administration (USA)
PIMS	pipeline integrity management systems
PSIM	pipeline structural integrity management
QFTA	quantitative fault tree analysis
QRA	quantitative risk assessment (or analysis)
RAM	reliability, availability, and maintainability
RAW	risk achievement worth
RBI	risk-based inspection
RCM	reliability-centered maintenance
ROI	return on investment analysis
RRW	risk reduction worth
SCC	stress corrosion cracking
SOP	standard operating procedure
SRB	sulfate-reducing bacteria

Author details

Marco Antônio Sabará[1,2]*, Jose Antônio da Cunha Ponciano Gomes[3] and
Alysson Helton Santos Bueno[1]

1 Federal University of São João del Rei – UFSJ, São João del Rei, Minas Gerais, Brazil

2 Federal Institute of Minas Gerais – IFMG, Congonhas, Minas Gerais, Brazil

3 Federal University of Rio de Janeiro – UFRJ, Rio de Janeiro, Rio de Janeiro, Brazil

*Address all correspondence to: marco.sabara@ifmg.edu.br

References

[1] Omoya OA, Papadopoulou KA, Lou E. Reliability engineering application to pipeline design. International Journal of Quality & Reliability Management. 2019; **36**(9):1644-1662

[2] Faisal K, Rioshar Y, Rouzbeh A. Risk-based pipeline integrity management: A road map for the resilient pipelines. Journal of Pipeline Science and Engineering. 2021;**1**(1):74-87

[3] Kishawy HA, Gabbar HA. Review of pipeline integrity management practices. International Journal of Pressure Vessels and Piping. 2010;**87**(7):373-380

[4] Mingjiang X, Zhigang T. A review on pipeline integrity management utilizing in-line inspection data. Engineering Failure Analysis. 2018;**92**:222-239

[5] Umair S et al. Enhancing pipeline integrity: A comprehensive review of deep learning-enabled finite element analysis for stress corrosion cracking prediction. Engineering Applications of Computational Fluid Mechanics. 2024;**18** (1):2302906

[6] Sani AA et al. Integrating life cycle cost analysis into pipeline asset integrity management: A comprehensive approach in decision support systems. Journal of Hunan University Natural Sciences. 2024;**51**(1):91-111

[7] Noshin AL et al. A quantitative approach of measuring sustainability risk in pipeline infrastructure systems. Sustainability. 2023;**15**(19):14229

[8] Hemalata J et al. Corrosion protection practices and integrity management challenges in oil and gas pipeline. In: Transportation Systems Technology and Integrated Management. Singapore: Springer Nature Singapore; 2023. pp. 141-162

[9] Pankaj T. Practical application of AI to transform pipeline integrity management. In: AMPP Corrosion. AMPP; 2023. pp. 39-59. AMPP-2023-19039

[10] Jiatong L et al. Data modeling techniques for pipeline integrity assessment: A state-of-the-art survey. IEEE Transactions on Instrumentation and Measurement. 2023;**72**:1-17

[11] Guojin Q et al. Failure probability estimation of natural gas pipelines due to hydrogen embrittlement using an improved fuzzy fault tree approach. Journal of Cleaner Production. 2024;**448**: 141601

[12] Xiufeng Z, Jinyu A. A new pre-assessment model for failure-probability-based-planning by neural network. Journal of Loss Prevention in the Process Industries. 2023;**81**:104908

[13] Cong W, Chunni W. Research progress of failure mode diagnosis methods for oil and gas field pipelines. Academic Journal of Science and Technology. 2024;**9**(1):192-197

[14] Hussein FHM, El Amine Ben SM, Tarek Z. A comprehensive review of corrosion protection and control techniques for metallic pipelines. Engineering Failure Analysis. 2023;**143**: 106885

[15] International Electrotechnical Commission. IEC 61025: Fault Tree Analysis (FTA). Geneva: IEC; 2006

[16] Wang Y, Xie M, Xu X. Optimal maintenance strategy for natural gas pipeline systems under uncertain conditions. Available at SSRN 4825156

[17] Chun L, Jinshi L, Di Z. Fuzzy fault tree analysis and safety countermeasures

for coal mine ground gas transportation system. PRO. 2024;**12**(2):344

[18] Chengtao W, Guojin Q. Corrosion of underground infrastructures under metro-induced stray current: A review. Corrosion. Communications. 2024

[19] Qunfang H et al. Unraveling mechanisms and failures of stray current corrosion in urban Subway adjacent pipeline networks through statistical insights and 3D simulation. Science of the Total Environment. 2024;**927**:172407

[20] Xiaolong L et al. Numerical simulation of stray current corrosion of buried steel pipelines. In: Fourth International Conference on Mechanical, Electronics, and Electrical and Automation Control (METMS 2024). SPIE; 2024. pp. 881-887

[21] Liu H et al. Effect of stray current on corrosion behavior of Mg alloy sacrificial anode in buried pipeline. Engineering Failure Analysis. 2023;**143**:106852

[22] Moran AJ, Lillard RS. A Modeling approach to understanding the interrelated nature of Cathodic protection current and AC stray current on pipelines. Corrosion. 2023;**79**(5):526-538

[23] Bingyuan H et al. Dynamic Bayesian network risk probability evolution for third-party damage of natural gas pipelines. Applied Energy. 2023;**333**:120620

[24] Haonan M et al. Advances in corrosion growth modeling for oil and gas pipelines: A review. Process Safety and Environmental Protection. 2023;**171**:71-86

[25] Vishnuvardhan S, Murthy A, Ramachandra CA. A review on pipeline failures, defects in pipelines and their assessment and fatigue life prediction methods. International Journal of Pressure Vessels and Piping. 2023;**201**:104853

[26] Jie Y et al. Leak detection and localization techniques in oil and gas pipeline: A bibliometric and systematic review. Engineering Failure Analysis. 2023;**146**:107060

[27] Hongfang L, Dongmin X, Guojin Q. Environmental risk of oil pipeline accidents. Science of the Total Environment. 2023;**874**:162386

[28] Deepankar C, Haque CC. A critical review on performance of buried pipeline subjected to pipe bursting and earthquake induced permanent ground deformation. Soil Dynamics and Earthquake Engineering. 2023;**173**:108152

[29] Rui X et al. Understanding the factors and consequences of pipeline incidents: An analysis of gas transmission pipelines in the US. Engineering Failure Analysis. 2023;**152**:107498

[30] Tanmoy S, Sutra DA. Lateral ground movement effects near a joint of MDPE gas distribution pipes. Canadian Geotechnical Journal. 2023;**60**(9):1415-1428

[31] Aikaterini T, Dimitrios L, Panos P. Design of offshore gas pipelines against active tectonic fault movement. Soil Dynamics and Earthquake Engineering. 2023;**166**:107702

[32] Ghadeer M et al. Internal corrosion in oil and gas wells during casings and tubing: Challenges and opportunities of corrosion inhibitors. Journal of the Taiwan Institute of Chemical Engineers. 2023;**150**:105027

[33] Uyen D et al. Modeling and analysis of internal corrosion induced failure of oil and gas pipelines. Reliability Engineering & System Safety. 2023;**234**: 109170

[34] Hussein KA et al. Emerging AI technologies for corrosion monitoring in oil and gas industry: A comprehensive review. Engineering Failure Analysis. 2023;**155**:107735

[35] Vinooth R et al. Corrosion monitoring at the interface using sensors and advanced sensing materials: Methods, challenges and opportunities. Corrosion Engineering, Science and Technology. 2023;**58**(3):281-321

[36] Zebei R et al. Predicting the external corrosion rate of buried pipelines using a novel soft Modeling technique. Applied Sciences. 2024;**14**(12):5120

[37] Li YT et al. Prediction model for failure pressure of large-diameter gas pipelines at different temperatures. Journal of Pipeline Systems Engineering and Practice. 2024;**15**(2):04024015

[38] Changzhou W, Lijing Z, Gang T. Quantifying the influence of corrosion defects on the failure prediction of natural gas pipelines using generalized extreme value distribution (GEVD) model and copula function with a case study. Emergency Management Science and Technology. 2024;**4**(1):1-6

[39] Muhammad W, Djukic MB. External corrosion of oil and gas pipelines: A review of failure mechanisms and predictive preventions. Journal of Natural Gas Science and Engineering. 2022;**100**:104467

[40] Yun-Tao L, Xiao-Ning H, Shuai J. Risk analysis and maintenance decision making of natural gas pipelines with external corrosion based on Bayesian network. Petroleum Science. 2022;**19**(3): 1250-1261

[41] Pan J et al. Investigation into the mechano-electrochemical interaction of internal and external corrosion defects on pipe surfaces. International Journal of Pressure Vessels and Piping. 2024;**207**: 105098

[42] Rui X et al. Predicting failure pressure of corroded gas pipelines: A data-driven approach using machine learning. Process Safety and Environmental Protection. 2024

[43] Bingqin W et al. Identification of corrosion factors in blast furnace gas pipe network with corrosion big data online monitoring technology. Corrosion Science. 2024;**230**:111906

[44] Jia Z et al. Safety and reliability assessment of external corrosion defects assessment of buried pipelines—Soil interface: A mechanisms and FE study. Journal of Loss Prevention in the Process Industries. 2023;**82**:105006

[45] Weifan C et al. Inspection of Liner Wall thinning and Interface Debonding in bimetallic lined pipes using pulsed Eddy current testing. Materials. 2024;**17** (11):2652

[46] Niantao Z et al. A prediction method of wellhead movement with consideration of interface debonding. Geoenergy Science and Engineering. 2024;**239**:212919

[47] Hamed S, Reg E, Weixing C. A review on current understanding of pipeline circumferential stress corrosion cracking in near-neutral PH environment. Engineering Failure Analysis. 2023;**148**:107215

[48] Kaiyang L, Yimin Z. Long-term corrosion and stress corrosion cracking

of X65 steel in H_2O-saturated supercritical CO_2 with SO_2 and O_2 impurities. Construction and Building Materials. 2023;**362**:129746

[49] Xiaoguang H et al. The synergistic effect of temperature, H_2S/CO_2 partial pressure and stress toward corrosion of X80 pipeline steel. Engineering Failure Analysis. 2023;**146**:107079

[50] Thibankumar A et al. The influence of axial compressive stress and internal pressure on a pipeline network: A review. Applied Sciences. 2023;**13**(6):3799

[51] González-Velázquez JL et al. Fracture mechanics analysis of the stress corrosion cracking failure of stainless steel hexagonal head screws in a marine-industrial environment. Engineering Failure Analysis. 2023;**146**:107098

[52] Alaso JT, Jerzy S. Effect of Nb-induced microstructure on pipeline steel corrosion and stress corrosion cracking performance in acidic environment. Corrosion Science. 2023;**218**:111196

[53] Zhong L et al. Stress corrosion cracking failure of X80 carbon steel U-bend caused by Desulfovibrio vulgaris biocorrosion. Journal of Materials Science & Technology. 2024;**174**:95-105

[54] Dongxu S et al. Study on stress corrosion behavior and mechanism of X70 pipeline steel with the combined action of sulfate-reducing bacteria and constant load. Corrosion Science. 2023;**213**:110968

[55] Olha Z et al. Application of the various methods for assessment of in-service degradation of pipeline steel. Mechanics of Advanced Materials and Structures. 2023;**30**(24):5058-5067

[56] Wenhe W et al. Experimental investigation of stress corrosion on supercritical CO_2 transportation pipelines against leakage for CCUS applications. Energy Reports. 2023;**9**:266-276

[57] Ziyue H et al. A dynamic condition assessment model of aging subsea pipelines subject to corrosion-fatigue degradation. Applied Ocean research. 2023;**139**:103717

[58] Xinhong L, Mengmeng G, Guoming C. A hybrid algorithm for inspection planning of subsea pipelines subject to corrosion-fatigue degradation. Process Safety and Environmental Protection. 2023;**178**:685-694

[59] Leila F, Mohtadi-Bonab MA. A review on the corrosion and fatigue failure of gas turbines. Metals. 2023;**13**(4):701

[60] Ziyue H, Xinhong L, Guoming C. A stochastic model for RUL prediction of subsea pipeline subject to corrosion-fatigue degradation. Process Safety and Environmental Protection. 2023;**178**:739-747

[61] Yusheng Y et al. Corrosion fatigue behavior of X65 pipeline steel welded joints prepared by CMT/GMAW backing process. Corrosion Science. 2023;**225**:111568

[62] Olha Z et al. Evaluating hydrogen embrittlement susceptibility of operated natural gas pipeline steel intended for hydrogen service. Engineering Failure Analysis. 2024;**163**:108472

[63] Maximilian R et al. A review of gas phase inhibition of gaseous hydrogen embrittlement in pipeline steels. International Journal of Hydrogen Energy. 2024;**60**:1239-1265

[64] Leonardo G et al. Embrittlement, degradation, and loss prevention of

hydrogen pipelines. MRS Bulletin. 2024; **49**(5):464-477

[65] Xiusai X et al. Experimental study on the temperature dependence of gaseous hydrogen permeation and hydrogen embrittlement susceptibility of X52 pipeline steel. Engineering Failure Analysis. 2024;**155**:107746

[66] Shiyao P et al. Influence of hydrogen volume/specimen surface area ratio on hydrogen embrittlement sensitivity of X52 pipeline steel. International Journal of Pressure Vessels and Piping. 2024; **209**:105217

[67] Hongchao Y et al. Effects of defect on the hydrogen embrittlement behavior of X80 pipeline steel in hydrogen-blended natural gas environments. International Journal of Hydrogen Energy. 2024;**58**:158-173

[68] Cailin W et al. Inhibiting effect of carbon monoxide on gaseous hydrogen embrittlement of pipelines transporting hydrogen. Corrosion Science. 2024;**227**: 111789

[69] Aashna R et al. Evaluating hydrogen gas transport in pipelines: Current state of numerical and experimental methodologies. International Journal of Hydrogen Energy. 2024;**67**:136-149

[70] Chengshuang Z, Hongbin Z, Lin Z. The impact of impurity gases on the hydrogen embrittlement behavior of pipeline steel in high-pressure H_2 environments. Materials. 2024;**17**(9): 2157

[71] Associação Brasileira de Normas Técnicas. NBR ISO 15589-2:2012: Indústrias de petróleo, petroquímica e gás natural — Proteção catódica de sistemas de transporte por dutos — Parte 2: Dutos submarinos. Rio de Janeiro: ABNT; 2012

[72] Xi C, He R, Cees B. Fault tree analysis for composite structural damages. Proceedings of the Institution of Mechanical Engineers, Part G: Journal of Aerospace Engineering. 2014;**228**(9): 1466-1474

[73] Shen Y, Zhou W. A comparison of onshore oil and gas transmission pipeline incident statistics in Canada and the United States. International Journal of Critical Infrastructure Protection. 2024; **45**:100679

[74] Greg L, Mumbi M-H, Pekney NJ. Underground natural gas storage facility operations and well leakage events in the United States. Geoenergy Science and Engineering. 2024;**234**:212630

[75] Silva BF, De Oliveira EVA, Cunha DA, Goncalves M. Change of location class in gas pipelines from a regulatory perspective. Process Safety Progress. 2024;**43**(2):313-320

[76] Pfander I et al. Extensive pipeline location data resource: Integrating reported incidents, past environmental loadings, and potential geohazards for integrity evaluations in the US Gulf of Mexico. Data in Brief. 2024;**55**:110728

[77] Ahmad MD et al. Collection and analysis of hydrocarbon gas buried onshore pipeline accidents in Indonesia as the databases for failure frequency assessment in a quantitative risk analysis. Process Safety Progress. 2024; **43**:S128-S133

[78] Larive JF. Performance of European cross-country oil pipelines. In: Statistical Summary of Reported Spillages in 2006 and since 1971. 2008

[79] Baker M Jr, Fessler RR. Pipeline corrosion (final report) US Department of Transportation. In: Pipeline and Hazardous Materials Safety

Administration, Office of Pipeline Safety, Integrity Management Program under Delivery Order DTRS56-02-D-70036. 2008

[80] Incidents, Gas Pipeline. 11th Report of the European Gas Pipeline Incident Data Group (Period 1970–2019). Group. IGU. Milan; 2020

[81] Mcconnell RA, Haswell JV. Ukopa pipeline product loss incidents (1962–2010). In: 8th Report of the UKOPA Fault Database Management Group, UKOPA/11/0076. Vol. 2011. Ambergate: UK Onshore Pipeline Operators' Association; November 2011

[82] Ernest L et al. Statistical Review and Gap Analysis of LNG Failure Rate Table. 2017

[83] Dawotola AW et al. Estimation of failure rates of crude product pipelines. In: Köhler & Nishijima, editors. Applications of Statistics and Probability in Civil Engineering–Faber. London: Taylor & Francis Group; 2011

[84] Lianshuang D et al. Analysis and comparison of long-distance pipeline failures. Journal of Petroleum Engineering. 2017;**2017**(1):3174636

[85] Golovach YA, Kornilova NV. Method development of the integral specific failure rate determination considering time in service and diameter of gas pipelines. IOP Conference Series: Materials Science and Engineering. 2020; **773**:012065

[86] Cunha SB. Comparison and analysis of pipeline failure statistics. In: International Pipeline Conference. American Society of Mechanical Engineers; 2012. pp. 521-530

[87] Anjuman S, Rehan S, Solomon T. Risk analysis for oil & gas pipelines: A sustainability assessment approach using fuzzy based bow-tie analysis. Journal of Loss Prevention in the Process Industries. 2012;**25**(3):505-523

[88] Caleyo F et al. On the Estimation of Failure Rates of Multiple Pipeline Systems. 2008

[89] Dong Y, Yu D. Estimation of failure probability of oil and gas transmission pipelines by fuzzy fault tree analysis. Journal of Loss Prevention in the Process Industries. 2005;**18**(2):83-88

[90] Canadian Energy Pipeline Association. Pipeline Industry Performance Report, Calgary, Canada. 2015

[91] PHMSA. Pipeline Incident 20 Year Trends. 2024. Available from: https://www.phmsa.dot.gov/data-and-statistics/pipeline/pipeline-incident-20-year-trends [Accessed: June 24, 2024]

[92] Mohsen A, Reza BM. A comparative reliability study of corroded pipelines based on Monte Carlo simulation and Latin hypercube sampling methods. International Journal of Pressure Vessels and Piping. 2020;**181**:104079

[93] Keonhee P et al. Comprehensive framework for underground pipeline management with reliability and cost factors using Monte Carlo simulation. Journal of Loss Prevention in the Process Industries. 2020;**63**:104035

[94] Zaki S, Yuri L. Risk analysis of an underground gas storage facility using a physics-based system performance model and Monte Carlo simulation. Reliability Engineering & System Safety. 2020;**199**:106792

[95] Xinfang Z et al. Reliability-based assessment of cracked pipelines using Monte Carlo simulation technique with

CorLAS™. In: Pressure Vessels and Piping Conference. American Society of Mechanical Engineers; 2022. p. V002T03A051

[96] Weichao Y et al. Subset simulation-based reliability analysis of the corroding natural gas pipeline. Reliability Engineering & System Safety. 2021;**213**: 107661

[97] Hongfang L et al. An inventory of greenhouse gas emissions due to natural gas pipeline incidents in the United States and Canada from 1980s to 2021. Scientific Data. 2023;**10**(1):282

[98] Kimiya Z, Fuzhan N, Tarek Z. Availability-based reliability-centered maintenance planning for gas transmission pipelines. International Journal of Pressure Vessels and Piping. 2020;**183**:104105

[99] Bhardwaj U, Teixeira AP, Soares CG. Uncertainty in reliability of thick high strength pipelines with corrosion defects subjected to internal pressure. International Journal of Pressure Vessels and Piping. 2020;**188**:104170

[100] Long C et al. Bayesian Monte Carlo simulation–driven approach for construction schedule risk inference. Journal of Management in Engineering. 2021;**37**(2):04020115

[101] International Organization for Standards. Uncertainty of Measurement-Part 3: Guide to the Expression of uncertainty in measurement. ISO/IEC GUIDE 98-3:2008(E). Geneva, Switzerland; 2008

[102] Jun L, Yan Mingqing Y, Jianjun. Evaluation on gas supply reliability of urban gas pipeline network. Eksploatacja i Niezawodność. 2018;**20**(3):471-477

[103] Feng L et al. Analysis on accident-causing factors of urban buried gas pipeline network by combining DEMATEL, ISM and BN methods. Journal of Loss Prevention in the Process Industries. 2019;**61**:49-57

[104] Ahmed SA et al. Integrity assessment of corroded oil and gas pipelines using machine learning: A systematic review. Engineering Failure Analysis. 2022;**131**:105810

[105] Haoran H, Baikang Z, Jian G. Natural gas pipeline leakage analysis based on advance risk analysis and analytic hierarchy process. IOP Conference Series: Earth and Environmental Science. 2020;**1427**: 012019

[106] Kimiya Z, Fuzhan N, Tarek Z. A review of failure prediction models for oil and gas pipelines. Journal of Pipeline Systems Engineering and Practice. 2020; **11**(1):03119001

[107] Biezma MV et al. Most fatal oil & gas pipeline accidents through history: A lessons learned approach. Engineering Failure Analysis. 2020;**110**:104446

[108] Aihua L et al. Dynamic risk assessment model of buried gas pipelines based on system dynamics. Reliability Engineering & System Safety. 2021;**208**: 107326

[109] Zhonghong H, Jian S. Performance evaluation method of oil and gas pipeline integrity management. Journal of Loss Prevention in the Process Industries. 2023;**84**:105099

[110] Alan M. Pipeline emergency response protocols and incident investigation. In: Handbook of Pipeline Engineering. Cham: Springer International Publishing; 2023. pp. 1-32

Chapter 5

Environmental Impact of Built Structures during Their Lifecycles in Terms of Construction Waste

Renata Androšević

Abstract

Construction industry waste has received much attention in recent years, as it produces more than a third of the total waste. A huge expansion of structures that are still built in the same way amplifies this negative environmental impact. Research was conducted with the mission of finding a solution and finding an efficient method that can ease the decision-making process in designing and building low-waste buildings. The final goal is to have a circular building economy where all building parts can be reused in each phase of its first and any next lifecycle. The method and Construction Waste Elimination Protocol was developed in the doctoral research in 2022 by the author of this chapter, and she wants this updated version to be available for broader visibility to speed up its digitalization.

Keywords: building industry, environmental impact, reducing construction waste, reuse methodology, building life cycle management

1. Introduction

Due to the fact that the construction industry has a dramatic increase in quantities of built structures and, according to that, the significant impact in a global environmental pollution, it is of great concern to find the powerful tool to lower the negative impact. Environmental pollution manifests in many aspects, like climate change, resource depletion, water, soil, and air pollution, extinction of wildlife, etc. If we succeed in minimizing and stopping the negative consequences of building in at least one of the categories, the quality of life will be improved.

The purpose of the research was to describe the current problem with construction waste, examine the possibilities of lowering the quantities, and establish the concept of "no waste building." No waste building refers to the practice of designing, constructing, and operating buildings in a way that minimizes their environmental impact, conserves resources, and promotes the well-being of occupants. Even though there are numerous theories on how to build a building like that, it is necessary to better align theoretical research with practical applications, consequently leading to more sustainable and environmentally friendly practices.

The basis of the study is the innovative initiatives with aim to develop a sustainable built environment that minimizes harm to nature and all living beings. The research seeks to aid stakeholders in making informed decisions by identifying materials, building technologies, and systems that have the least detrimental environmental impact, while providing essential data for consideration.

> *"Construction waste is identified as a problem of 21st century and this research will go one step forward in solving this issue...The desirable objective of the research is to contribute to the design and implementation of more sustainable and reusable buildings and building systems to lower the building waste, as one of the problems in total building environmental impact" [1].*

By examining current methods and tools, the study looks to address the challenges in achieving sustainable construction practices. Existing methodologies underscore the significance of predicting environmental impacts and allow for comparisons between different systems, while tools offer precise numerical data on environmental effects. It is suggested that the combined use of methodologies and tools can lead to the desired outcomes of minimal waste in construction practices.

Therefore, the proposed approach involves evaluating building systems through comparison and ranking, along with quantifying their environmental impact. This process gives a comprehensive understanding of a system's effect and aids in decision-making. Ultimately, the goal is to identify areas for improvement and to enhance performance, while reducing environmental harm.

Through this new method, comparisons, evaluations, and quantifications regarding construction waste generation are made to minimize negative environmental impacts of built structures. The research suggests areas for further research and potential improvements to the methodology and provides recommendations for practitioners in the construction industry to implement low-waste building practices.

2. Methodology in scientific approach while creating the protocol

The research started with the overview of the growing environmental concerns related to impact that construction brings to environment and the need for sustainable practices.

Environmental impact of built structures, with a specific emphasis on construction waste, was detected as one of the major problems to solve, as it has a significant role in overall environmental degradation. "Based on volume, construction and demolition waste is the largest waste stream in the EU" [2]. European construction and demolition waste is equivalent to around 46% of the total amount of waste generated [3]. Data shows voluminous streams worldwide, The European Commission aims to increase confidence in the waste management process, establishing measures to reduce the volume of construction demolition waste sent to landfills, conserve natural resources, and lower negative environmental impact.

Root cause analysis investigates the expansive growth of built structures and its correlation with increased waste production. Recently, an increasing number of construction and demolition waste management projects have been implemented. On the other hand, the building industry continues to build in the same way, and new potential construction demolition waste will be produced repeatedly. Little work has been done to prevent the waste. According to the European Commission, the first and preferred option is waste prevention, and the second option is other types of recovery (like preparing for reuse, recycling, recovery, and disposal as the least preferred option) [4]. There is, evidently, an urgent need to look ahead, predict scenarios, and enhance the reusability of all building parts to avoid waste in the first place.

Solution convergence explores existing strategies and innovations aimed at reducing construction waste, such as theories and concepts of low- or no-waste buildings. Environmental philosophies applied to building design show historical evolution from "green design," "eco design" to "sustainable design" building, and finally, "design for sustainability," where radical design is striving toward a sustainable future [5]. World Green Building Council adopted the Agenda for Sustainable Development, and one of the adopted Sustainable Development Goals (SDG) was "Responsible consumption and production." When we apply that to the building industry, it means that we have green buildings that use "circular" principles, where resources are not wasted [6]. Considering all mentioned, low-waste sustainable buildings should be designed with a focus on material and energy efficiency, with predicted scenarios and strategies for deconstruction and material reuse at the end of each lifecycle. The goal is to include all parts of the building in a circular low-waste building industry and circular economy [1]. There is an urgent need to transfer from a "linear" to a "circular" concept (**Figure 1**).

Considering all of the above, the research project has developed a new method – a systematic approach to designing and constructing buildings with minimal or no waste in each lifecycle. Total building waste consists of building production and construction waste, building maintenance waste and demolition waste [1]. "Circular" building model means that waste is avoided in every phase in the lifecycle of the building project and the building itself. The phases in the building lifecycle are design, realization/production/construction, use/maintenance, transformation/reuse, and deconstruction/demolition (in the worst-case scenario) [7].

A building will be decomposed into systems, which will be tested separately to identify the method and the tool that give the best evaluation [1] and establish the Protocol that supports it. Construction waste, which plays a key role in total

Traditional linear concept "take – make - dispose" with the waste at the landfill.

New concept "take – make – use again" that forms loop.

Figure 1.
Two points of view of the phases in each project [1].

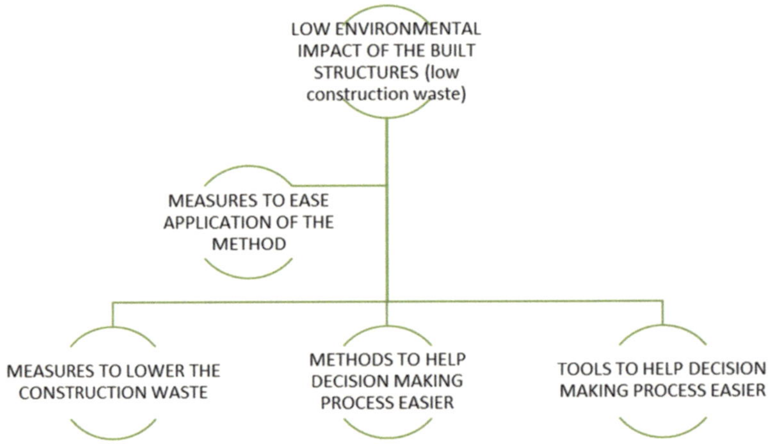

Figure 2.
Establishing construction waste prevention [1].

environmental impact, will be decreased when following measures, methods and tools to lower construction waste in design phase (**Figure 2**).

The expression "buildings as material banks" [8] can describe this "circular construction" where resources can be reduced, and materials can be reused until the end of their technical lifecycle and after that period, simply recycled and brought back in the loop again.

After the profonde research, Construction Waste Elimination (CWE) Protocol was created [1]. With the ground base of this Protocol, a decision-making tool should be made to facilitate the design and construction of low-waste buildings. Protocol was tested through chosen case studies and representative buildings to apply the new Protocol.

Among the first case study, the decomposition of the building into systems was made. A comparison of the results from different case studies and different methods was used to identify the most effective solutions. Evaluation assesses the performance of the method and tool in real-world scenarios and measure outcomes against predefined criteria. Despite considerable research, translating theoretical knowledge into effective, practical solutions remains a challenge. Inevitably, focusing on these areas, the construction and building industry can better align practical applications with theoretical research, ultimately leading to more sustainable and environmentally friendly practices.

3. Research integration

For the purpose of the research, the first case studies were selected on the basis of use of the new building technology, place of design (Bosnia and Herzegovina), and territory where it was built. Three case studies were tested for their reuse potential using the method Durmišević [1, 9]. The reuse potential at the building level and unit level gave different results. With the intention to achieve more accurate results, it is necessary to test every building system separately.

The next step was to choose the representative building system to test the existing methodology and tools. When discussing the longevity of buildings and their components, it is important to consider the varying durability rates of these building parts. Durability of services is 7 years, partitioning 10 years, enclosure 20 years, and frame approximately 70 years, as outlined by Durmišević [9]. Data analysis and assumptions point out that the main goal for the new building system proposal and its testing should be enclosure [1]. BIH is a country very rich with wood and the old settlements were made mostly from wood and stone [1], so the wooden wall was chosen as a representative building system. Wooden wall systems are:

a. Energy efficient,

b. Easy for assembly and

c. Affordable [1].

Benefits and sustainability advantages of wood structures were demonstrated in real time case studies. Wood is highly preferable material, as proved in "Light Wood-Framed Residential Structures" thanks for their "design for deconstruction" [10].

Two wooden walls were selected as case studies. Both walls have wood as the primary material, and both are previsioned as walls of energy-efficient buildings. One wall is a wall for prefabricated energy efficient house (**Figures 3** and **4**), and the other wall is for energy efficient sustainable transformable building Green Design Center (GDC) (**Figures 5** and **6**) [1].

In the mentioned research, both walls were tested for reuse potential and used as case studies to compare the other methods and tools' competency in the field of environmental impact and construction waste. Comparison of the reuse potential of the two case study walls underlines the importance of environmental impact calculation in the early phase of the building project. In the decision-making process for this study [1], considering its specific context, the following databases, methods, and tools have been selected to measure and predict the environmental impact of built structures, and they can assist in forecasting construction waste:

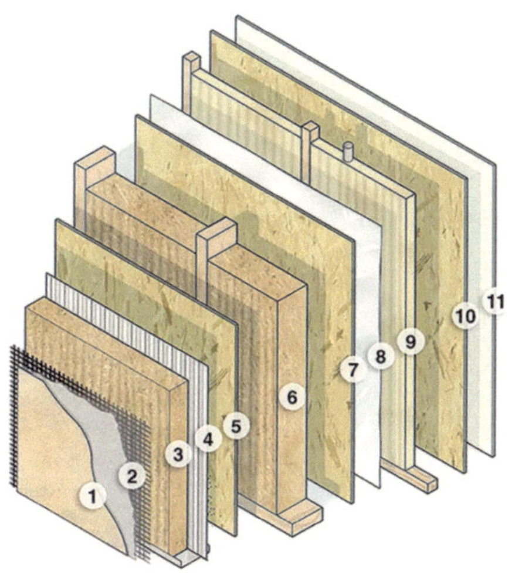

1. Decorative plaster finish 2,0 mm
2. Reinforced mortar and mash 4,0 mm
3. Thermal and sound insulation:
Wood fiber insulation 100,0 mm
4. Dispersion adhesive
5. OSB 3 Board 12,0 mm
6. Wooden load-bearing construction
Thermal and sound insulation Glass Mineral Wool 160,0 mm
7. OSB 3 Board 12,0 mm
8. PE foil - vapor barrier 0,2 mm
9. Insulated service cavity Thermal and sound insulation Glass Mineral Wool 50,0 mm
10. OSB 3 Board 12,0 mm
11. Gypsum plasterboard 12,5 mm

Figure 3.
Krivaja 11 layers' low energy house wall [11].

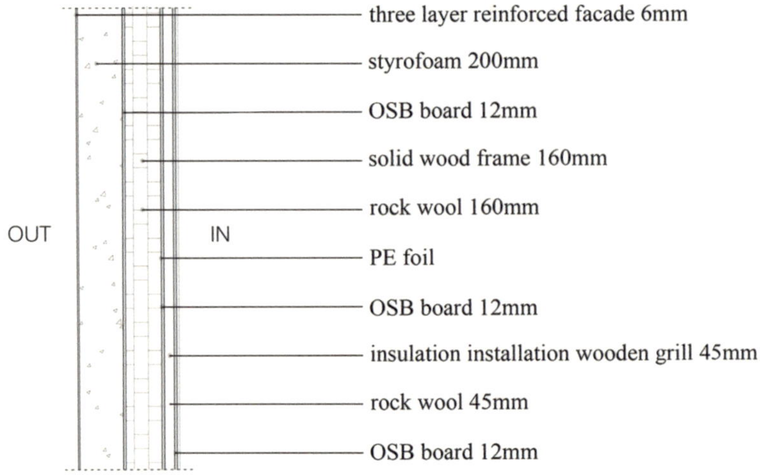

Figure 4.
Krivaja 11 layers' low energy house wall layers updated [1].

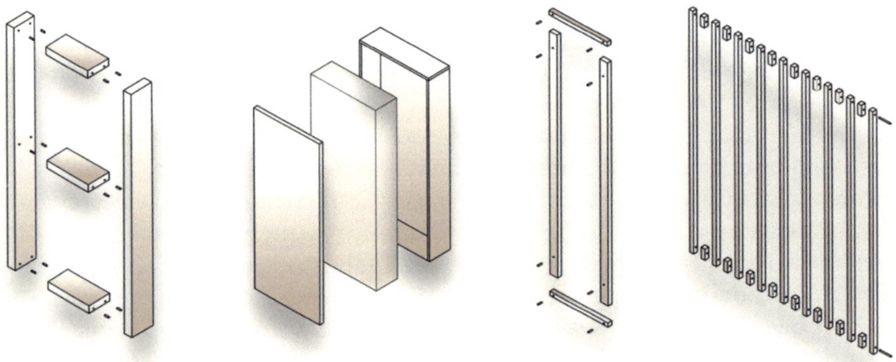

Figure 5.
GDC low energy wall layers (a) main frame of the system, (b) insulation box, (c) intermediary, (d) finishing of the system [12].

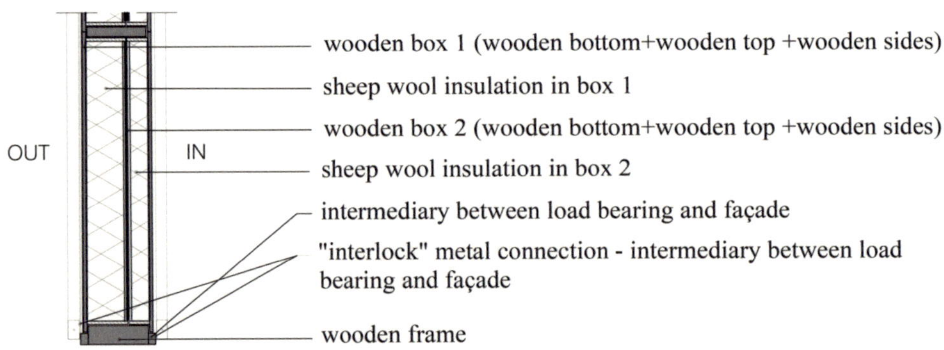

Figure 6.
GDC low energy wall layers [1].

a. Building materials databases (MD)—helpful to track the origin and composition of substances/materials in a product/system.

b. Building Information Modeling (BMI) which can use the material databases—helpful to show the composition of materials in the system and their relations.

c. Reversible building design (RBD) protocols which include method for measuring the reuse potential (RP) of a building.

d. Methods and tools for analyzing and measuring the impact—Life Cycle Assessment (LCA) tools.

All these databases, methods and tools were studied and investigated separately in order to find the best solution to help in the decision-making process in the building industry in the field of environmental impact and construction waste [1]. Recapitulation of advantages and disadvantages of these databases, methods and tools, according to the research Androšević [1], are in the **Table 1**.

In attempt to reach the goal of sustainability, the upgrading of the existing LCA software should enable the synthesis with the following:

- database of connection materials,
- building parts relations,
- scenarios of possible transformations and upgrade in the first lifecycle,

Method/ tool	**Advantages/helps to lower construction waste and negative (EI)**	**Disadvantages/missing information**	**Suggestions for adequate:**
BMI	• Opportunity to plan, design and see the building in its entirety with its parts. • Provide the possibility to work on the same model and upgrade it with different functions. • All materials, components, elements and systems with their quantities in one place.	• Missing materials, because of different file formats of MD. • Not suggesting sustainable details. • Depends on the knowledge of project designer.	• BIM platform: connection with other tools (RBD and LCA) that suggest behaviors and possible scenarios with calculations for EI.
RBD	Rp indicator shows level of reusability. (Higher Rp = lower construction waste and vice versa)	Not connected with material database, missing quantities of construction waste and total EI.	• RBD tool: connection with material database, BIM and LCA tool.
LCA	Connected with material database and shows EI for each material and calculates total EI.	Missing material relations (how to separate materials to reuse them)	LCA tool: connection with more MD, than BIM and RBD tool.

Source: Adapted [1].

Table 1.
Advantages and deficiencies of methods and tools that help in project sustainability and lower negative environmental impact (EI).

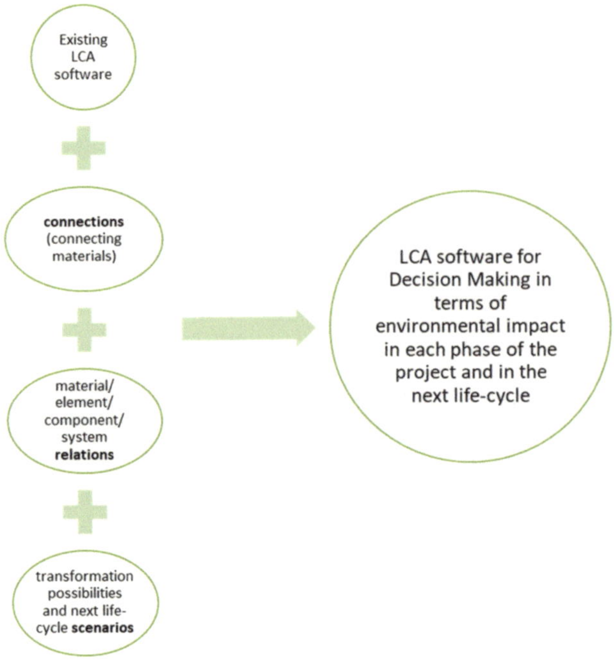

Figure 7.
Suggested upgrade for the LCA software which would help in decision process regarding the environmental impact of the future building [1].

- scenarios of possible transformations in the next lifecycles (**Figure 7**) [1].

The integrated method which considers all mentioned recommendations, develops into the Construction Waste Elimination (CWE) protocol [1].

4. Construction waste elimination (CWE) protocol

According to the conducted research (1), the CWE protocol was created, and here will be presented the summary of that process. The new CWE protocol is envisioned as an instruction to follow in order to achieve low construction waste in each phase of the building life (design, realization/production/construction, use/maintenance, transformation/reuse, deconstruction/demolition), and of course, in each lifecycle [1]. To improve building design concerning construction waste and minimize environmental impact, the decision-making process should begin with defining and predicting the time frame. While many buildings are envisioned to last indefinitely, in practice, the lifespan and evolution of a building often come down to these two broad scenarios:

a. The building will remain in its original state and serve its intended function for at least 50 years or more. Buildings that are designed to remain in their original state and serve their intended function often have characteristics like timeless design, durable materials, and a purpose that does not change over time. These structures are built with the assumption that their use and esthetic will remain relevant and functional for decades.

b. The building will undergo functional changes and experience significant transformations within the next 10–20 years. Conversely, buildings that are expected to undergo functional changes and transformations often have to reflect more dynamic needs and uses.

Both scenarios highlight the importance of considering future adaptability in building design, whether it is through robust, timeless construction or flexible, change-ready features. In both scenarios, there will be a need to modify building systems for maintenance, upgrades, and repairs. When proposing/predicting potential use scenarios for a building in a requested time frame, the possible timeline can be [1]:

a. Shorter than the expiry date of the material/part with the shortest technical lifetime,

b. The same as the expiry date of the material/part with the shortest technical lifetime,

c. Longer than the expiry date of the material/part with the shortest technical lifetime.

It is important to predict possible scenarios for each building system (in this case, the external wall). For each scenario, the reuse potential should be measured for parts and the whole system. If the system is declared with "high reuse potential" (Reuse potential (Rp) value higher than 0.6 [13]), low construction waste creation with low environmental impact is expected. It is a basic framework and gives directions to stakeholders when deciding what system to choose to have possibilities to repair, reuse, and upgrade the system and its parts (some of them separately, some of them in the system) within different transformations or maintenance purposes in the same lifecycle or for some other purpose in another lifecycle [1]. Number of transformations and upgrade possibilities should be predicted and visible to the stakeholders, as these are system's value preservations during time. Buildings and their parts should be designed for transformation and multiple use.

The CWE protocol should highlight the differences in potential waste and its environmental impact across the lifecycles of various case studies in building planning. The average Conn value indicates the part of the element, component or system that cannot be reused in subsequent lifecycles, which will be discussed as a key factor contributing to waste generation after the building's first lifecycle (**Table 2**).

If the predictions are accurate, the building technology should facilitate the upgrade, extraction, and reuse of systems, components, elements, and materials, contributing to a circular and sustainable building industry.

The integrated method will outline steps for each phase of the building process:

a. predesign/design,

b. realization/production/construction,

c. use/maintenance,

d. transformation/reuse,

e. deconstruction/demolition [1].

	First lifecycle '		Second lifecycle "	
	Average conn value	Environmental impact/EI/per chosen category	Environmental impact/EI/per chosen category	Environmental impact/EI/per chosen category
Case Study 1 (CS1)	(Conn = 1-Rp)	EI1'	EI1'	EI1"
Case Study 2 (CS1)	Conn1'	EI2'	EI2'	EI2"

Source: Adapted [1].

Table 2.
The CWE protocol illustrates the differences in potential waste and its environmental impact throughout the lifecycles of various case studies in building planning. The Conn value indicates the portion of the product that cannot be reused, while Rp represents the calculated Reuse Potential.

a. After predesign phase (where building parts have the proof and warranty that they will last and be maintained for predicted timeframe, under certain condition), there comes design phase, which considering the scenarios, specific calculations and necessary drawings required by Law, according to the new methodology. The construction phase cannot (should not) start before the next four steps in design phase:

   a. First step—material relations (the network of relationships and connections between building components should offer a clear understanding of their connections and the possibilities for assembly and disassembly).

   b. Second step—reuse potential calculations with the focus on relational dependencies, type of connections, and assembly and disassembly possibilities [9].

   c. Third step—bill of materials which includes connecting materials and their relations (upgrade for the initial bill of quantities required by Law), preferably with material "expiry date"—done by hand or BIM tool connected with MD.

   d. Fourth step—environmental impact (EI) calculation in the LCA software, considering all inputs from the previous steps [1].

According to the research [1], the EI in terms of construction waste quantities for the first building lifecycle (until the first transformation) for the entire building or for a specific part of the building, can be simply calculated using the LCA software by selecting the appropriate methodology. The EI for the second lifecycle of the building or its parts can be calculated considering reuse potential and connections between parts.

These data play a significant in the total EI score. It is necessary to calculate the total EI in terms of Construction Waste Elimination (CWE) in the second lifecycle. The total value depends on the relations between parts and not just on the quantities of used materials and their individual impact. As explained in LCA tools, One Click LCA and Open LCA, it is assumed that the impact of all materials (in the production and use phase) is calculated in the first lifecycle and if used in the second lifecycle, their impact is calculated as 0. In this research, the LCA tool is used to calculate the building material impact on the environment and does not consider other impact categories.

"There is necessity to introduce the Conn x-y values to underline distinction between the parts which can be disassembled and the ones which cannot. Conn values are assigned to the connection types used between building parts or materials. Value depends on the disassembly possibilities. The Conn values can be explained depending on the type of the connection proposed by Durmišević" [9].

Reuse potential connection indicator is determined, and its value is maximal; therefore, Rp is 1. If Rp has maximal, then Conn value, which represents waste after the disassembly, can be nearly 0. According to that, the Conn value would be the waste material that cannot be reused after disassembly and can be calculated for other connection types, according to the equation:

$$\text{Conn} = 1 - Rp \quad (1)$$

b. When construction begins at a site, there will inevitably be materials and products that are not used in the building and packaging that ends up as waste.

Effective building planning and accurate calculations can help minimize this on-site waste following next recommendations:

- preparation and prefabrication with a precise bill of quantities, especially the bill of materials.
- reducing the amount of packaging and
- using biodegradable materials.

As prefabricated buildings have been abandoned from designers for a while (with explanation of uniformity without creativity), combination of prefabrication with on-site construction will be preferred solution. When on-site modifications are necessary, producers should encourage the return of both leftover materials and product packaging, to support circular building practices and reduce waste [1].

c. Leveraging digitalization during the use and maintenance phase is highly recommended, including mapping all installed materials once the building is completed. Creating a comprehensive database of materials used in the building (within the specified timeframe) is essential. Each component should be traceable for future material tracking and potential harvesting. Database of built-in material include the following information for each building component:

a. the installation date,

b. its expected technical lifespan, and

c. its relational pattern for potential transformation or end-of-life scenarios [1].

d. The transformation phase is smoother and more cost-effective if transformations were predicted in design phase. When building components are well-documented and their value is preserved over time, transformation becomes straightforward and economical. A crucial element in this process is a database that tracks reusable parts. This database is essential for achieving sustainable building practices. This approach will result in:

a. Minimal environmental impact,

b. Preservation of economic value, and

c. Enhanced social value [1].

e. If the previous steps are followed as outlined, the deconstruction phase should be straightforward and valuable. This phase should focus on "deconstruction" rather than "demolition," with the goal of minimizing waste by maximizing material reuse. Deconstruction can occur at the end of the building's first lifecycle or during phases requiring part replacements. When materials are easily accessible, and disassembly is well-planned, stakeholders can repurpose building components for new projects once the building's initial lifecycle is complete. For maintenance-related deconstruction, ease of disassembly ensures that part replacement and reassembly are quick. If deconstruction is anticipated, all components should be documented in a reuse database, including information such as the deconstruction date and material availability. With proper planning and information flow, materials can be transported directly to the new construction site, bypassing the need for temporary storage.

CWE protocol was tested on GDC wall case study [1] and gave good view on the possibilities of the materials, elements, components and systems of the wall in the following lifecycles. CWE framework should be integrated with the tool that facilitates the decision process. Even though it is recommended to have the tool "all in one," it should consider different options for different stakeholders, as different stakeholders are interested in different impact calculations. Some different options should be considered for the next groups of stakeholders:

a. project designers (architecture/product design/consultancy company)

b. building constructors (building parts producer/construction/demolition

c. company)

d. building investors (private or public) [1].

5. Discussion and conclusion

The significance of new environmental protocols globally is evident, and waste prevention is crucial issue in environmental protection. The research Androšević [1] established a comprehensive overview of existing assessment methods and tools for measuring building and building parts environmental impact, as a design framework for buildings and their systems, components, and elements that aims to minimize environmental impact in every lifecycle. This framework enhances understanding of how design solutions influence future transformation and reuse possibilities, and consequentially influence total environmental impact. It offers an objective assessment of compared case studies and supports circular building practices.

The construction waste is not just a simple quantity of the waste, for example weight (kg) or volume (m3) of all parts and materials included. It depends on:

a. the characteristics of the used materials,

b. material relations (connections),

c. material condition after dismantling.

d. of the maintenance and conditions during building parts lifetime,

e. of the total time passed while in use [1].

The construction waste quality and quantity and its environmental impact depend on several factors that should be carefully considered, including:

a. Relational dependencies between parts of the system,

b. Type of the connections,

c. Assembly possibilities,

d. Disassembly possibilities (all four are defined by Durmišević [9]),

e. Possibility of building transformations – calculated only for the whole building, and not necessarily for the building system,

f. Quantity of the total materials used,

g. Quantity of the materials/parts that can be reused (with expiry date),

h. Quantity of the materials/parts that can be recycled (with zero or positive balance),

i. Time frame for calculation (predicted timeline and expected lifetime of the parts),

j. Number of transformations [1].

The Construction Waste Elimination (CWE) protocol provides a precise picture of the analyzed building or building product concerning construction waste and its impact. Integrating the construction industry into a circular economy does not entail sacrificing profit. It requires a shift in mindset and adherence to new methodologies, such as the Construction Waste Elimination protocol, along with careful planning [1]. It is possible to lower the construction waste quantities, if we follow the proposed Construction Waste Elimination protocol, which consists of steps that include combination of existing methods with some upgrades.

The Laws should add new regulations, specify and define the "transformable building," which should and could follow the change in function, in accordance with the user needs. The future building structure must allow change and transformation, in order to be included in circular construction and economy, which means that the building can:

a. be able to change and upgrade its parts,

b. accept and change parts with the other building,

c. be easily disassembled and all its parts can be the parts in another building lifecycle,

d. transform the shape during its lifecycles [1].

Considering the above, new construction/building Construction Waste Elimination protocol must be considered as the requirement for the building permit. The set of documents in architectural building project must have a depository of all used materials/elements/components/systems in a building and their relations besides the set of drawings, calculations and other building technical information.

Successful reduction of construction waste is a complex process, and its' environmental impact is a result of next factors' combination:

a. Construction companies adopt a new mindset and revise their approaches.

b. Waste generation is meticulously quantified and tracked.

c. Stakeholders select appropriate construction methods, employ skilled and well-trained workers, and ensure thorough design supervision [14].

d. Prefabricated building elements are utilized off-site in precast designs.

e. Decision-makers proactively predict construction waste amounts and establish benchmarks to reduce waste generation [15].

f. Designing out waste becomes a primary objective for architects and contractors within the framework of sustainable construction [16].

g. The building system with the lowest waste generation is chosen.

h. building is defined as transformable structure (and not permanent!),

i. environmental impact of a building is defined by law and low environmental impact of a building is obligatory for a building permit,

j. low waste building becomes the investor's main concern. (Encouragement could start when the cost of disposing waste/transportation costs and landfill/become so high that) [1].

Limitations that the methodology is facing are following:

a. Partially acceptance of prefabrication from stakeholders (mostly designers),

b. Unreliable data from producers,

c. Difficult validation of the available building product data [1].

Future work should be done in the direction of upgrading the material database and LCA tools, adding:

a. data for the materials that are connections (the connection materials, connection possibilities, and relations between building parts should be in focus),

b. sustainable suggestions for relations between materials, systems, and products,

c. possible transformation and upgrade scenario during the first lifecycle, reliable plan for dismantling the building or its parts,

d. traceability of all the parts of the building in all lifecycles [1].

In the future, Reversible Building Information Modeling (BIM) [17] will have a significant role at the beginning of the architectural project, when the connection and synergy with other software and material databases are achieved (to lower the environmental impact). Interconnection and compatibility of material databases, building information modeling (BIM), reversible building design (RBD), and life-cycle analysis (LCA) tools should be obligatory. Only under these conditions can the data and results be clear, allowing for the correction of current and future flaws and the prevention of negative environmental impact [1].

The preliminary results after testing one building system, demonstrate the feasibility of the methodology and Construction Waste Elimination (CWE) protocol. The next step will be to apply this methodology to other building systems and entire buildings. Based on these tests, the methodology should be refined and upgraded to address more complex systems.

Cooperation with software engineers with the goal to digitalize methodological framework and address the existing gaps is essential. The compatible building information modeling (BIM), reuse potential (RP) and lifecycle analyses (LCA) software is vital in digitalization of the whole process. It can be “the implementation of Industrialized Building Systems (IBS), which has the potential of promoting sustainability deliverables” [18]. Digitalization ensures quicker results, easier track and flow of information and transparent data sharing for all stakeholders [1].

Once the proposed changes to the database and software are implemented, it will become easier to develop and test complex building systems during the early design phase. Although there is still a significant path ahead for the methodology’s acceptance and digitalization, stakeholders are likely to overcome these challenges as the methodology framework gains broader acceptance.

The adoption and development of green building protocols are crucial as they help to ensure that new structures not only meet the growing needs of the population but also contribute positively to environmental stewardship and sustainability. As these principles become more refined and widely adopted, they will play a significant role in shaping the future of construction and building practices.

Nomenclature

Acronyms

BIM	building information modeling
CWE	construction waste elimination

EI	environmental impact
EU	European Union
GDC	green design center
IBS	industrialized building systems
LCA	lifecycle analyses
MD	materials databases
RBD	reversible building design
RP	reuse potential
SDG	sustainable development goals
Symbols	
Conn	value that indicates the portion of the product that cannot be reused

Author details

Renata Androšević
Faculty of Architecture, University of Sarajevo, Sarajevo, Bosnia and Herzegovina

*Address all correspondence to: renata.androsevic@af.unsa.ba

References

[1] Androšević R. Going toward low waste external walls of residential buildings in XXI century in Bosnia and Herzegovina [thesis]. Ville: Transports et Territoires, Paris Est Sup University; 2022

[2] European Commission (EC). EU Construction and Demolition Waste Protocol and Guidelines. 2018. Available from: http://ec.europa.eu/growth/content/eu-constructionand-demolition-waste-protocol-0_en [Accessed: February 25, 2021]

[3] Gálvez-Martos JL, Styles D, Schoenberger H, Zeschmar-Lahl B. Construction and demolition waste best management practice in Europe. Resources, Conservation and Recycling. 2018;**136**:166-178. DOI: 10.1016/j.resconrec.2018.04.016. ISSN 0921-3449

[4] European Commission (EC). European Parliament and Council. Directive 2008/98/EC of the European Parliament and of the Council of 19 November 2008 on Waste and Repealing Certain Directives. Official Journal of the European Union L 312/3. 2008. Available from: http://data.europa.eu/eli/dir/2008/98/oj [Accessed: September 20, 2024]

[5] Moreno M, De los Rios C, Rowe Z, Charnley F. A conceptual framework for circular design. Sustainability. 2016;**8**(9):937. DOI: 10.3390/su8090937

[6] Tokede OO, Rodgers G, Waschl B, Salter J, Ashraf M. Harmonising life cycle sustainability thinking in material substitution for buildings. Resources, Conservation and Recycling. 2022;**185**:106468. DOI: 10.1016/j.resconrec.2022.106468. ISSN 0921-3449

[7] Androšević R, Durmišević E, Brocato M. Measuring reuse potential and waste creation of wooden façades. In: IOP Conference Series: Earth and Environmental Science. Vol. 225, No. 1. England: IOP Publishing; 2019. p. 012017. DOI: 10.1088/1755-1315/225/1/012017

[8] Debacker W, Manshoven S. D1 Synthesis of the State-of-the-Art. Key Barriers and Opportunities for Material Passports and Reversible Design in the Current System (H2020 BAMB Project). 2016. Available from: https://www.bamb2020.eu/wpcontent/uploads/2016/03/D1_Synthesis-report-on-State-of-theart_20161129_FINAL.pdf [Accessed: September 20, 2024]

[9] Durmišević E. Transformable Building Structures: Design for Disassembly as a Way to Introduce Sustainable Engineering to Building Design and Construction. Delft: Delft University of Technology; 2006. Available from: https://resolver.tudelft.nl/uuid:9d2406e5-0cce-4788-8ee0-c19cbf38ea9a [Accessed: August 5, 2024]

[10] Guy B, Shell S, Esherick H. Design for deconstruction and materials reuse. Proceedings of the CIB Task Group. 2006;**39**(4):189-209

[11] Krivajahomes. 2020. Available from: www.krivajahomes.com [Accessed: September 20, 2024]

[12] Brussels Environment (Groens, H., Capelle, T., Henrotay, C., Steinlage, M.)—authors of the report. UTwente (Durmišević, E., Beurskens, P.), EPEA (Mul, E.J.), SGDF (Androšević, R.), VUB (Paduart, A., Elden, S., Lanckriet, W., Poppe, J.)—authors of the pilot project studies. D13 Prototyping and Feedback Report—Testing BAMB Results through Prototyping and Pilot Projects. EU Horizon 2020 Project BAMB Work Package 4. 2018. Available from:

https://www.bamb2020.eu/wpcontent/uploads/2018/10/20180425-BAMB-WP4-D13.pdf [Accessed: September 20, 2024]

[13] Durmišević E. Circular Economy in Construction Design Strategies for Reversible Buildings. The Netherlands: University of Twente; 2019. ISBN 978-90-821-698-4-3

[14] Scheer S, de Amorim SRL, Santos ET, Ferreira RC, Caron AM. The scenario and trends in the Brazilian IT construction applications experience. Journal of Information Technology in Construction (ITcon). 2007;**12**(13):193-206

[15] Kern AP, Dias MF, Kulakowski MP, Gomes LP. Waste generated in high-rise buildings construction: A quantification model based on statistical multiple regression. Waste Management. 2015;**39**:35-44. DOI: 10.1016/j.wasman.2015.01.043

[16] Osmani M. Design waste mapping: A project life cycle approach. In: Proceedings of the Institution of Civil Engineers - Waste and Resource Management. Vol. 166, No. 3. England: ICE Publishing; 2013. pp. 114-127. DOI: 10.1680/warm.13.00013

[17] Durmišević E, Guerriero A, Boje C, Domange B, Bosch G. Development of a conceptual digital deconstruction platform with integrated reversible BIM to aid decision making and facilitate a circular economy. In: Proceedings of the Joint Conference. CIB W78-LDAC. Vol. 2021. Luxembourg; 2021. pp. 11-15

[18] Yunus R, Yang J. Sustainability criteria for industrialised building systems (IBS) in Malaysia. Procedia Engineering. 2011;**14**:1590-1598. DOI: 10.1016/j.proeng.2011.07.200